SELECTED P

SELECTED POEMS

1972–1990

Peter Scupham

Oxford New York
OXFORD UNIVERSITY PRESS
1990

Oxford University Press, Walton Street, Oxford OX2 6DP
Oxford New York Toronto
Delhi Bombay Calcutta Madras Karachi
Petaling Jaya Singapore Hong Kong Tokyo
Nairobi Dar es Salaam Cape Town
Melbourne Auckland
and associated companies in
Berlin Ibadan

Oxford is a trade mark of Oxford University Press

This Selection first published in Oxford Poets
as an Oxford University Press paperback 1990

British Library Cataloguing in Publication Data
Scupham, Peter, 1933–
Selected Poems 1972–1990.—(Oxford poets). 2
I. Title
821.914
ISBN 0–19–282762–6

Library of Congress Cataloging in Publication Data
Data available

Typeset by Wyvern Typesetting Ltd
Printed in Great Britain by
J. W. Arrowsmith Ltd, Bristol

Leave the door open, let the landing through
To stop the room from being itself tonight,
Part of the rain, the leaves, the outer darkness
At odds with the particulars of sight.

Leave the door open, for the air is close:
A kind of heat rubbed on a kind of fear.
Aimless things are about some urgent business,
The safe and far is dangerous and near.

Leave the door open; though the kitchen voices
Climb and fall back, all they can ever say
Is that they have no spell, there is no meaning,
Nothing is further off than break of day.

Leave the door open, let one golden bar
Sink a cool shaft from ceiling down to floor.
Light: steady and simple, an assurance
To make the door's assurance doubly sure.

'Leave the Door Open', from *Out Late*

Acknowledgements

The Snowing Globe was published by E. J. Morten (Publishers) in 1972, as one of the Peterloo Poets Series edited by Harry Chambers. The other seven books were published by Oxford University Press.

Contents

From *The Snowing Globe*, 1972
Address Unknown 1
Four Fish 2
The Snowing Globe 3
Children Dancing 4
The Occasion 7
Painted Shells 8
Unpicked 11
The Nondescript 13

From *Prehistories*, 1975
Ploughland 14
Prehistories 16
Wind and Absence 18
Divers 19
As If 20
Birthday Triptych 21
Grass Dancers 23
Vertigo 24
No Cause 25
Christmas Lantern 26

From *The Hinterland*, 1977
The Spring Wind 27
Atlantic 28
Trencrom 30
The Hinterland 31
Effacements 39
Marginalia:
1 Leaf from a French Bible, *c.* 1270 41
2 Jeremy Taylor: The Rule and Exercise of Holy Dying, 1663 42
3 Rider's British Merlin, 1778 43
Answers to Correspondents 44
Under the Words 45
Arena 45

From *Summer Palaces*, 1980
Twelfth Night 47
Three Sisters:
1 The Dry Tree 48
2 Wild Grass 49
3 The Curving Shore 50
Horace Moule at Fordington 51
Summer Palaces 52
The Beach 53
The Chain 54
The Gatehouse 55

From *Winter Quarters*, 1983
The Spanish Train 56
from Notes from a War Diary:
Madelon 57
Echoes 58
Fin 59
from Conscriptions: National Service '52–'54:
Scapegoat 61
Prisoner 62
Sentries 63
Possessions:
1 (M. R. James) 64
2 (Walter de la Mare) 65
3 (Rudyard Kipling) 66
La Primavera 67
from Transformation Scenes:
Scott's Grotto, Ware 69
Ornamental Hermits 70
The Makers 71
Kingfisher 72
The Candles 73

From *Out Late*, 1986
A Borderland 75
A Midsummer Night's Dream:
1 Prologue 76
2 Wind 77
3 The Elementals 77
4 Shadows 78
5 The Court 79
6 The Green World 80

7 The Lovers 80
8 Constellations 81
9 The Mechanicals 82
10 Moon 83
11 Epilogue 83
Green 85
A House of Geraniums 86
The Key 87
Genio Loci 88
The Gardens 89
Ragtime 90
Cat's Cradle 91
from Erotion:
Martial V, 34 92
Martial V, 37 92
Martial X, 61 93

From *The Air Show*, 1988
Under the Barrage 94
The Other Side of the Hill 95
Diet 96
Bigness on the Side of Good 97
The Old Frighteners 98
Double 99
Going Out: Lancasters, 1944 100
The Wooden-heads 101
His Face 102
V.E. Day 103
Jacob's Ladder 104
Blackout 105
Service 106
Magic Lantern 107

From *Watching the Perseids*, 1990
The Master Builder 108
Christmas 1987 110
Young Ghost 111
Dancing Shoes 112
Watching the Perseids: Remembering the Dead 113

Address Unknown

Who goes roaming at night through my bricks and mortar?
What tremors wander through my patchwork rafters?
Why do these small holes drip a slight buff powder?

I think of bulky mammals who roved here stolidly,
Munching and pulling at the ancient greenery.
A million years. I could sleep them away tonight.

My house is clipped lightly to an old hillside,
Held against cloud and shine by the wills of others,
Tacked up with sticky grit and threads of power.

Its teeth are set on edge by a guitar trembling;
Each thought unpeels a slim skin from the paintwork.
House, I have stuffed you with such lovely nonsense:

All these sweet things: Clare's Poems, Roman glassware,
A peacock's feather, a handful of weird children,
A second-hand cat with one pad missing, missing.

Where are you going, carrying us without spilling,
Sailing away, packed for a hero's funeral,
Freighted with love-gifts, all destined for strangers?

What is not there at all is what lasts longest,
The airs, the hill it seems that we have borrowed,
A hill itself settling to a new calm level.

Old tigers and stuff poke through my dirty windows.
At the door, Doctor Who swishes an antic mantle,
Commanding the bleeping future to throw me sideways.

I lie in bed; pack the house into my headroom.
Though bits hang out, it is safe from hags and urchins.
I shall put a bone round it, and a spell of closing

To keep away the Spoom, the Man in the Oke, the Succubi
Who shake small tremors through my patchwork body,
Rocking me awake with changeling, manic voices.

Four Fish

Gay chalks and a blackboard. Watch Kate trace
Four fish. They hover softly in dark space,

Warming the kitchen by their tranquil glow,
Infallible, heraldic. Pictures flow

With casual extravagance, to live
Assured as flowers, and as fugitive.

These lack all claim or pretext; merely say:
Here is a world, and it behaves this way.

They mock our making by their candour; lift
Our hesitant, blunt senses; are true gift

For Rilke's Angels. Knowing our task is
To feed the invisible with our images,

We rub the blackboard clear: four fish allow
To swim in darker space more subtly now.

The Snowing Globe

A trick of the fingers
And the world turns.
A drift of tiny snow.

The frosted reindeer,
Lost in a Christmas wood—
They will make room for us.

Thicker, colder the airs.
Softly, neatly,
You hurry to where I wait.

Your rough black coat
Brushes the sidling flakes.
They cling like burrs.

Under an ancient light,
Snow in your dark hair
Dissolves; glistens like tears.

Our eyes take fire.
We are preserved in amber
These fourteen crystal years.

Somewhere, our unborn children
Smile with your smile.
The snow is settling now.

Children Dancing

for Margaret

1

What shall we throw down
To image again the one theme,
Link our games to the one game?

The ache of print in stiff books,
The wrinkled vanities each day employs,
Our distances, our reservations.

What shall we piece together
To image again the one theme,
Link our games to the one game?

These fragile bones, built from dust,
A tinsel shade, a skein of music,
Our laughter, our serious eyes.

Now we accept all broken steps,
Provisional identities;
Wait on the eyes' absorption

Till patterns flow to the one pattern,
Delay Time's hatchet-men,
Her separations and habitual snows.

The elemental and undying children
Shift soft colours on a tolerant darkness.
We warm ourselves at their unwearying fires.

2

The cracks in an old floor
Crinkle across my eyes.
I feel threadbare.

How vulnerable she is:
I know her sweats, her tremors.
Admiration distresses me.

Having unlearnt such candour,
Must I become Prometheus,
Stealing fire from a skimpy body

Whose sweet, unsteady flesh,
Trailed hair, spaced fingers,
Are close and far as first love?

Supple in a damp leotard,
Her painted face beaten by light,
She is that Roman horn

Ever spilling, ever full.
Through her, our words and music
Assume strong presences.

A child moves towards death,
Shaping with each slight gesture
Ignorant benedictions.

3

Shadow and exhaustion overtake us.
After occasions of sunburst
All movement concurs toward silence.

Rigged frames, floods, battens,
Wear a shut, flimsy look,
Slewed to the blank flats.

Can I thread the dancers back
On to your voice of rough silver,
The sprung energies of music?

No. Laughing with linked peers,
They dismantle their borrowed faces,
Scatter our laborious archives.

We have our photographs:
Butterflies, slowly stiffening
In aromas of crushed laurel.

Rather, let memory play out
Her own ring dances and mandalas;
Compose cool, brilliant landscapes

Where we and the lost children can rehearse
Our affirmations, our secret codes,
Across the drifting years.

The Occasion

The cups and glasses started to unfold
Warm gestures against different sorts of cold.

The girl was dark and very luminous.
She was herself, and that, for sure, was precious.

The air was mild, the air was calm and tender:
Her bones were tricksy, and her wrist was slender.

The man read poems in a handsome way,
Their shapes now somewhat grave, now somewhat gay.

The lissom wife let fingers trip and stir
At her guitar. Softly, he looked at her;

Moved by the thoughtful stranger who lay hid
In that familiar sweet flesh and blood.

Smiles walked on tiptoe: bright or lazy birds
Which settled on the different sorts of words.

Some glasses brim so full of mortal light,
They make a Jack O'Lantern of the night.

Oh, strong amazements time is careless of.
Oh world, who spares her creatures little love.

The girl was dark; her hushed, observant eyes
Spun the room widdershins, back again clockwise.

She was a brilliant fluttering in a ring.
She was, he thought, a quick, precarious thing.

The curtains came down softly on this play
When glasses and guitar conspired to say:

Alone, on different sorts of roads, we go,
Through sheets of flowers and through sheets of snow.

Painted Shells

Children pick painted shells to hear
Long tides come hushing on the ear;
Believe that in half-understood
And mortal rumours of their blood
Live groundswells of a deeper sea:
A language of Eternity.
Unmoved by flattery or command,
Our brutal hour-glass slips her sand
Where tides and tokens all concur
That child and dancer must defer
To Time, whose known inclemency
Will freeze the small bird from the tree,
And whose vindictive whiplash gives
No quarter to her fugitives.

Whatever cancers wait, we must
Confirm our hearts' instructions; trust
We are not only what we know
Our savagery and folly show:
Mere vacancies the dust assumes
To link dull cradles to dull tombs,
Sick children crying for a light
Where sweats and nightmares wind us tight
And no one comes — and if they did
The face they'd show had best be hid.
Our presences the more enhance
The linking figures in our dance
When what we say and do creates
A bonfire to amaze the fates.

My dear, by this charm repossess
Those airs and fires your ways express:
A candid smile to lift each heart,
Green-fingered talent in that art
Which shows how bravely souls are dressed
And how we may reveal them best.
Whatever luck the years advance,
May your five senses lead their dance
With that intensity and verve
The frayed edge gives the steady nerve;

And may you turn those trolls to stone
Who try to claim you for their own,
Whose doltish needs reduce each grace
To a shut look on a shut face.

May the old world that lets you through
Kindly incline her fruits to you,
Her metal landscapes part to make
Paths for your dancing steps to take.
Carry through wild or temperate air
The dark declensions of your hair;
Make all hobgoblins, all annoyers,
Imps, black vanities, destroyers,
That shocking crew whose yammering flocks
Fled poor Pandora's gaping box
Cower before you. May your feet
Keep measure with the Infinite,
Held in that intricate design
By far more potent runes than mine.

No amulet of words can stay
Our tender structures from decay,
Though buds unfrosted yet by Time
May flower precariously in rhyme.
We learn to recognize and bless
Whatever tames our wilderness:
Truant, sweet, divergent creatures,
Passions dancing in their features.
Though Nature's laws are obdurate,
Her sentence we can mitigate:
The artist's work is paradox,
A Chinese puzzle which unlocks
To show those gifts that he was made,
Their brightnesses transposed, repaid.

In Time and out of Time we move,
Inflicting pain, inflicting love,
Crusoes who salvage what we can
In our short, haunted, patchwork span;
And still in every fragile shell
Curl labyrinths whose echoes tell
The dancing words they must begin
To trace again our origin,

That distant and delaying flood
Which courses under mortal blood.
Where rivers ease into the sea
All musics mingle. We shall be
Rocked in those long tides at the ear
Children pick painted shells to hear.

Unpicked

I am unpicked by every wandering thing:
That stormy girl, the impermanence of May,
The things birds tell me in the tunes they sing,
The tender tumult of my children's play.
All, all unpicks my bones, unpicks my bones;
I speak in dissonances, quarter tones.

I have flown thoughts and words so far down wind
No artifice will now decoy them home;
Hung out a heart for other hearts to find.
I am a man whom speech has driven dumb,
Crushed by the simple weight of being good.
I am a stone that brims with some dark blood.

I still have craft, and I can bandage you
The wounds of long-dead poets; in my car
These distant, careful hands know what to do.
I know small histories of beast and star.
I am the scattered pages of a book
Whose theme is lost. I wear an alien look.

I know the drowning fly, the shorn dark hair,
The soldier weeping on his iron bed
In some old hutment by a windy square.
I know the unhealing sutures in my head.
I am unpicked; I cannot face alone
The vigorous, careless damage I have done.

The soft bombs drumming on my childhood make
Their crazy thunder on my dreaming ear:
I know boys' savagery, and each mistake
That I accomplish draws me into fear.
I stand reproached and into danger where
The haunted faces of my friends lie bare.

There was a time I would have laughed to see
You bend and break under my wilful hand
In some wild contest, and your misery
Had fed those vanities I understand.
Such strength was worthless. A deep amnesty
Has called those weapons home that suited me.

I am unpicked, but if you need a way
Of making such destruction doubly sure,
A mocking word or glance in casual play
Will be sufficient. I shall still endure.
Rejection I have always understood:
Her enigmatic marble face, her hood.

You whom I chose, who seem to have a need
For what a scarecrow's pockets can supply,
Take what you want, if you have skill to feed
On simple food when other horns run dry.
I will keep little back. I hang my mail
In some museum on a withered nail.

I think of old-gold castles on French hills
And all those ancient slit and tippling towers,
Their deep-set, murderous, defensive skills
Open to soft invasions from the flowers.
Does each hurt bee die lacking power to sting?
I am unpicked by every wandering thing.

The Nondescript

I am plural. My intents are manifold:
I see through many eyes. I am fabulous.

I assimilate the suffering of monkeys:
Tiger and musk-ox are at my disposal.

My ritual is to swallow a pale meat
Prepared by my ignorant left hand.

It is my child's play to untie a frog,
Humble further the worm and dogfish.

When I comb the slow pond,
I shake out a scurf of tarnished silver;

When I steer the long ship to the stones,
A brown sickness laps at the cliff's foot.

Shreds of fur cling to my metalled roads,
Old plasters seeping a little blood.

I dress and powder the wide fields:
They undergo my purgatorial fires.

Come with me. I will shake the sky
And watch the ripe birds tumble.

It requires many deaths to ease
The deep cancer in my marrowbones.

I have prepared a stone inheritance.
It flourishes beneath my fertile tears.

Ploughland

They drudged across the season's palimpsest,
The feathered horses, driven into rain;
Their dapples and attendant shades impressed
Upon the skies which pulled them up the lane.
    Light moves about the knotted ploughland, sown
    With undistinguished seeds of chalk and bone.

A sullen field, pegged out with brittle trees,
A water-colour of the English School;
Her pawky hedgerows crumble to their knees,
The locked fires in her fractured flint lie cool.
    Old weathers broke her in, and we retrace
    The crows' feet etched on her dissolving face.

The lost teams and their leaders. All that care
Is mere commemoration by the wind.
A shower of bells freckles the cooling air;
Our booted feet drag with an awkward rind.
    Down in the levelled churchyard a few graves
    Remind the birds how well the past behaves,

Working along the bent flow of the land
With no more substance than our thought makes good:
A face limned on the air, a tenuous hand
To scribble in the margins of the wood.
    A rash of brickwork rubricates their claim:
    A foundered house, and a forgotten name.

The Home Farm lacked a tenant forty years,
Pigs littered grandly in the living room;
The rain is flighted with her children's tears,
A burnt tree's crater beckons to the tomb.
    One old man with his dog slows up the hill;
    A shooting party spaces out the chill

Where scatterings of green remind the day
Of beasts long cropped and felled: the molten dead
Still driving purpose through the common clay
In new returns of sacramental bread.
 The harnessed bells ride out upon their round,
 Tossing cold heads above the burdened ground.

Prehistories

1

Adrowse, my pen trailed on, and a voice spoke:
'Now, you must read us 'Belknap'.' My book was open.

I saw their faces; there were three of them,
Each with a certain brightness in her eyes.

I would read 'Belknap'. Then a gardener's shears
Snipped fatefully my running thread of discourse.

And in my indices, no poem upon which
I could confer this honorary title.

Foundering in dictionaries and gazetteers,
I came there: Belas Knap, a chambered tomb.

The lips are closed upon that withered barrow:
A dummy portal, a slant lintel hung

Beneath a scalp of ruinous grass, her walls
A packed mosaic of blurred syllables.

2

Entering is a deployment of small silences,
Frail collusions and participations.

A scrape of some sad traffic on the ear,
Bird song at her old insouciance,

I pull these down into an underworld
Of images alive in their dark shelter.

Such corridors are tacitly inscribed:
Do not abandon hope here, but desire.

When old men's voices stumble down the lane,
They too must be my dry accompaniment.

There is a shrinking in each new encounter,
A heaviness attendant on the work.

The vanished bone sings in her shallow alcove,
Making a sacred and astounding music.

Ghosts are a poet's working capital.
They hold their hands out from the further shore.

3

The spirit leans her bleak peninsulas:
Our granite words loosen towards the sea,

Or hold, by some wild artifice, at Land's End.
The Merry Maidens dance their come ye all,

Keeping time to the piper's cold slow-motion,
Their arms linked against the lichenous Sabbath.

And a menhir, sharp-set, walled about
By a dirt-farm, a shuffle of lean cows,

Accepts, as of right, our casual veneration,
All fertile ceremonies of birth and death.

I see you standing, the new life quick in you,
Poised on Chun Quoit into the flying sky.

There, in that grave the wind has harrowed clean,
Our children crouch, clenched in a fist of stone.

Wind and Absence

Down wind, down wind, a soft sweep of hours
Trawling in time. My pulse races into darkness.

Adrift, I draw your absence close about me;
Take the small ghosts of your hands to mine.

Your voice, your smile: such *son et lumière*.
My nerves conduct you round my floodlit bones.

Certain salt-water fools pester my eyes:
I stub them out with rough, dumb fingers.

The wind rehearses idle punishments,
Shaking a crush of unrecorded leaves.

A long-dead singer clears his ashen throat:
You creep his old cold music into life.

Poems bite on their pains. Dismayed, I know
The lines are scored in poker-work: black fire.

The haunted room trembles at your insistency.
Small, ghostly hands, allay the air's distress

As wind gathers, menacing our naked spaces
In dead languages of distance and rejection.

Divers

We knew infallibly: love marked the spot.
Oh, charted faces, time-worn and marine!

After a year tossed among waves of hair,
Brimmed eyes and flotsam glances,

In a free fall, we leave the surface play
And work of light; traverse the permeable deeps

That close about our origins. All births
Are dumbly recognized in these slow worlds

Where silence and the beautiful darknesses
Drift and brood above the spirit's dancing floor.

In echoing cadences of green slow-motion,
We celebrate the ancient mysteries:

Combers for lost gold in a lodge of sand,
Amazed discoverers of our buried selves.

As If

As if the dying chapel in her grove,
With a most inward look, turned a blank face
To sunset and the lucencies of parting.
Then, briefly, let her golden surfaces
Enrich the air with generated light.

As if the clouds became the darkening sky,
And one wild bush, trembling upon the night,
Shifted the airs about and roundabout,
True author of the rain's diverse commotion.

As if such native sleight-of-hand is yours,
Performed, though, to an astonished daylight.
Some gifts confuse all reciprocity.

Birthday Triptych

1

Birth days: as of the spirit.
She cannot dissemble.
We break a fiery bread.

How should we navigate
The waves' coarse turmoil
Without appointed stars?

A luminary day, then.
Light of the first magnitude,
Confirm our chosen course.

Tides, in their slow recession,
Delay about your fingers
A light sweet freight of shells.

When fresh seas break,
May that beached miraculous wrack
Still hold its water lights.

On this your name day,
Under Janus, God of thresholds,
Past and future both become you.

2

The true gift claims us.
Look! The flowered paper
Spills and crinkles.

A drift of white tissue:
Snow wreaths in May
We missed last winter.

And the small sign disclosed
Says in a new voice:
I tell of love in the world

To steady and delight you.
A long draught fills our horn;
We cannot exhaust her.

We read a common language
Runed in each offered hand.
Here, riddles are their answers.

Eyes rehearse tender cues
All images compose and celebrate
The selves we have become.

3

None can walk safely.
The roads are dark, unsigned,
The sky precariously blue.

We must endure
Flowers, the rain's refreshment,
Each beautiful absurdity:

Accept with clear laughter
The dissonance of white hair,
Pain at the source.

A tree shakes at your window
Her brilliance of leafwork,
Admiration heals us.

Our vulnerability preserves.
To counter such a strength
Time's tactics have no skill.

Love sustains. By this avowal
We cancel fear, whatever tremors
Approach us from the bowed horizon.

Grass Dancers

Old night, I know you at this quiet fabling,
Your crickets sprinkling their light vocables.

Life goes barefoot on this stubbled garden:
A hidden sea is talking in its sleep.

Beyond the lackwit stones, the granite faces,
A sisterhood of cows makes cavalcade.

Badgers possess this place; they come for alms,
Noble antique beggars in grey fringes.

I smile for you, you small ones at my feet,
Tying soft webs of trill from stem to stem.

We make no shadows in this august starlight;
We rhyme softly, each out on his limb.

Jimp saltatoria, brown grass dancers,
Little, they say, is known about your histories.

Male and female he created them. Sweet love.
The girls have ears. Only the boys can sing.

Vertigo

Weary of foundering in Victorian silence,
Where pale occasional lamps disperse the gloom,
We find the strength to make a slight transition
And pace the balcony beyond the room.

Hands on the elegant railing, we admire
The melancholic trees, the broken fence;
Below, the tussocks in the formal garden
Mourn for their topiaried lost innocence.

Then you, ingenuous, with cat-felicity,
Swing yourself lightly on the rocking bar,
Back to the lawns, your hands crossed negligently,
Conscious that certainty can lean so far.

I grip the rail, inconsequently know
The centres of our gravity misplaced,
While hands and hearts, in fair-ground convolutions,
Perform their ritual in measured haste.

The grasses hiss, and the lean wind displaces
Curled leaves that swiftly turn a madder brown.
I, stay-at-home, gaze where your body presses
The unrecalcitrant dark flotsam down.

But your inevitable fall delaying,
You tire, step delicately down, and stare
With mock solemnity, sly comprehension,
Through that deep emptiness of cheated air.

No Cause

No cause, no cause. Adrift, he felt limbs tremble,
His tight skin burn, and the untidy room,
Stunned by such rough assault, crouch watchful back,
A frigid sun dappling the chequered floor.

Far worlds away, the children's careless voices
Tripped and fell. Only the clock drove on
Its balanced monotone. In that harsh pause
She stood as numb as ice, the tears in check.

And all was alien, lost: the cup she held,
The dumb cat shifting on the sill
And the green garden-light beyond the door.
How could there be for this no cause, no cause?

Christmas Lantern

Such flim-flam: a hutch of crumbling card
And the light's composure. Still talisman,

I think of nights your lucencies beguiled,
While the house rocked to the soft blundering guns.

I set my heart upon your heraldries
Where snow lies faultless to a cobalt sky,

Counting your shaken stars, adrift on paths
Diminishing among your vivid pines.

Frail monstrance, with an old anticipation
I set you here above my sleeping children:

The branched reindeer at the huddled bridge,
The simple house, offering her candled windows.

For their eyes now, your most immaculate landscape
And all the coloured rituals of love.

The Spring Wind

The Spring wind blows the window grey and white,
Stripping the hangings, pouring out the rooms.
There are the dados, and the beaded moulding,
The gather-ye-roses on the children's wall.
Doors shake on their jambs: the spine of the house
Thrills as the sprung wood quivers, and goes still.
Something here high and cold, knowing old paper
And skirts of paint, flaking, and cream with age,
Grey floorboards, seamed and dry with mousy dust.
This loveless breath scours where a white face creases
And small hands knuckle over a yellow cot.
What calls is lonelier than wind, though wind is older
Knows more, and knows it yet more emptily.
Ah, how the air fingers the pastel walls,
The shrinking webs; how the old house is blown.
The tussocks bare their silver undersides;
Under the warm brocades, the wings are pale.

Atlantic

There's loss in the Atlantic sky
Smoking her course from sea to sea,
Whitening an absence, till the eye
Aches dazed above the mainmast tree.
    The fuchsia shakes her lanterns out;
    The stiffening winds must go about.

Green breakers pile across the moor
Whose frayed horizons ebb and flow;
Grass hisses where the garden floor
Pulls to a wicked undertow.
    A ragged Admiral of the Red
    Beats up and down the flowerbed.

Granite, unmoving and unmoved,
Rides rough-shod our peninsula
Where curlews wait to be reproved
And petals of a hedge-rose star
    The beaten path. Uncoloured rain
    Rattles her shrouds upon the pane.

Beyond the ledges of the foam
A dog seal sways an oilskin head;
The Carracks worked old luggers home,
Black rock commemorates the dead.
    The sea shouts nothing, and the shores
    Break to a tumult of applause.

But mildewed on the parlour wall,
The Thomas Coutts, East Indiaman,
Enters Bombay. Her mainsail haul
Swells to the light. A rajah sun
    Accepts her flying ribbon still,
    Though bracken darkens on the hill.

A peg-leg cricket limps the floor.
See, China Poll and Jack link hands.
Blown long ago on a lee-shore,
His hour-glass run on to the sands,
    He bids her wipe her eye, for soft,
    A cherub watches from aloft

Who knows our hulk is anchored fast,
Though timbers fret in their decline.
The cattle heads are overcast;
The gutter shakes her glittering line.
    Rage at the door. Winds twist and drown.
    We founder as the glass goes down.

Trencrom

The salt brushed pelt of trees could hide them:
Ogres and witches who play pitch and toss
Or loose an apronful of clumsy pebbles
To stun the landscape into graves and kitchens.
Their lives are long and legendary as bones,
Their sleep deeper and harder than our sleep.

Quieter the gods of estuary and sand,
Holding their smoky fingers to our lips
And easing middle distance into distance;
Bruising the grass to a low stain of brown,
Shimmering our goose-skin with mist fingers.

The dead men leave their sunken gates unposted.
An age of iron rusts into a silence
Made luminous, as evening light discloses
A cold intimacy we had not looked for.

As if even the stones could drop their veils,
Discovering some life under their scarred rind,
Or words form from a circumference of wind.

Rocks breeding chill in deep interstices.
The ferns darkening, and the lizards gone.

We are moving under, and beyond the hill.

The Hinterland

1

The summer opens where the days draw in.
Under the heat-haze, polished branches flock
To idle sound, and through a gauze-green skin
Cut sun lies lapidary on a book.
These are the instances a page retrieves:
A gravel walk, and a tall palisade,
The action of white light on the sharp leaves,
Tubbed oranges and lemons, placed and played.
The conversations of a house withdraw:
Birds walk among the foliage to uncover
Time's dark heart in the ash and sycamore
And all the summers counting lost things over.
    An urgent cipher through a wall of trees:
    Leaves pressing home their small advantages.

2

Leaves pressing home their small advantages
Beyond the sill, beyond the frayed sash-cords,
To this close darkness where the dead hold lease,
Their furniture impalpable as words.
From a green room they prompt the present play,
The ebb and flow of blood. Dust-crevices,
Where life is forced from an old sap. Each day
Our veins are charged with their impurities
This summer when the shagged elms tower and die.
Their played-out torches moulder in each hedge;
Maintain erect, under a changeless sky,
Their stumbled footholds and miscarriages.
    August, September, hang their weights upon
    The rim of summer, when great wars begin.

3

The rim of summer, when great wars begin
And Europe sprawls beneath her calvaries,
Licking the dust-bowl of a culture clean.
Ask for a rushlight, or a varlet phrase,
For woodcut figures cold as anchorites
In far cells of the mind? The dragon's teeth
Sown in a picture-book are Naseby fight.
What is the legacy these bones bequeath?
They were buried shallow, and the putrid meat
Soured through the grass, wept the green over
'So that the cattle were observed to eat
Those places very close for some years after.'
    We fat the dead for our epiphanies,
    But there's a no-man's-land where skull-talk goes.

4

But there's a no-man's-land where skull-talk goes:
She sits alone by the declining sea;
Winds turn continually upon their course
And all their circuits are a vanity.
To her slow heart-beat, frosted empires fail—
Bone-china bone, the watered eyes half-blind,
Dust hair. No loving palmistries can spell
The bird-runes on the stretched silk of her hand
Which knows the loosening of silver cords,
Breaking of golden bowls. For her, the dead
Are a thrown captain on a sepia horse,
An old man with a nonsense at his head:
    Culled shades hob-nobbing where the airs are thin,
    A hinterland to breed new summers in.

5

A hinterland to breed new summers in,
And an old August working in the bones
Of children barefoot in a Northern town
Where windows speak in a long monotone,
All their black waters running deep. The ways
To washstand, jug and basin are dark shut:
Light climbs up to a silver bar, and stays.
Our children freeze; their urchin gestures cut
To crowded scarecrows on a cobbled field.
Their capped heads and their runnelled hands are full
With all our violence, and the stone is held
Which cannot fall; flung out, a dull arc pulls
    The locked sun down the road where elms turn brown—
    The unfleshed dead refusing to lie down.

6

The unfleshed dead refusing to lie down:
The favoured ones, the ephemeridae
Who knew the enchantments of the lost domain
And made a switchback of the ancient hay,
Who played about the Rector's sleeves of lawn,
Who won the Craven or the Nightingale,
From whom the bright Edwardian summer shone,
For whom Death was the Keeper of the Wall:
'The sons of men, snared in an evil time',
Whose luck ran out, whose claim was disallowed—
The poplars all are felled: not spared, not one—
Who lies in France under a criss-cross shroud?
    A great space, and young voices echoing.
    What inch of sunlight gilds their vanishings?

7

What inch of sunlight gilds their vanishings,
The harvests of our crooked scythes and spades?
'Huge, statuesque, silent, but questioning'—
Nobody, on his casqued and white parade,
Who left a name, a card, some ribboned bronze,
A song uncaptured by anthologies,
Whose flesh was eaten 'till the very bones
Opened the secret garments of their cartilages,
Discovering their nakedness and sorrow.'
One with all spun and all the spinning leaves,
The sift of man held close in field and barrow.
The eyes confuse: that rank and file of graves,
    Those dragon's teeth, those nests the birds have flown,
    Those penitential litanies of stone.

8

Those penitential litanies of stone—
A prophylactic wreath lies bleaching there,
Washed by the sun, the interceding rain—
A granite finger to recruit each year
The dead, for whom two minutes can be spared
To make their speeches on the common ground,
To hear the feet withdraw, the hard flags bared,
The terrible trumpets for lost Rolands, wound
About the thinning heads of those who stand
Gathered and set apart, themselves the dead,
The citizens of the corrupted land,
The handful of old dust, the new seed-head,
    The may, the blackthorn, the November trees;
    A silence runs beneath these silences.

9

A silence runs beneath these silences
Where the shut churches founder in the green,
And all their darkness and their brightness is
One seamless robe to lap the creature in.
Bone-chamber, tumulus, the field of skulls:
The folded hands, and the high blackened helm,
Where honour and dishonour share the rolls
And night plays rook about the stiffening elms.
Here each arrival, each new harvesting,
Deepens the ripples in a strengthening pool:
The troubled fountain and corrupted spring
Find waters which must scarify to heal.
    Let the tree lie where the tree is felled,
    And there our conversations must be held.

10

And there our conversations must be held,
Behind the lettered spines, the faded boards,
Where a plain text of living is distilled
And old exemplars must correct new words,
Who saw the rose, sprung from its cloven hood,
Begin to put on darkness and decline,
Who knew the outworn faces and the weeds,
The vanity, the striving after wind;
Who spoke for those who set their hearts in print,
The pulped trees and the drab who hawked the rags,
The baffled cattle waiting to be skinned,
The blow-fly's roundel on the vellum page.
    Each book the substance and the shadow. Lives
    Where blood and stone proclaim their unities.

11

Where blood and stone proclaim their unities
Under the topsoil vagaries of green
Works the slow justle of the small debris:
The ruins of the Glacial Drift begin,
Driving the golden horns, the feathered shales
Over each crevice of the lost domain,
Making the way straight where the Blue Gault falls
To Greensand and her seams of Ironstone.
The skies are wind: spoil by a ruffled pool.
A perched farm blossoms on its ancestries.
Time's face exposed: the quickening muscles pull
Five fingertips across the centuries,
    Poised where the bedrock widens a dull gold
    And all the shadows cross on one high field.

12

And all the shadows cross on one high field:
A latter violence working in the grain,
The cockpit tangling on the shrouded hill,
The hangars closed, and the bent grass at noon
Twelve o'clock high. Bandit and Angel stand
And cool their wing-tips in a molten sun
Of Kentish fire, their long trails wound and fanned
Against a sky which cannot take the stain,
The lost crews working home against the wind,
Against new traffics and discoveries.
The elms are singled, failing one by one:
Earth entertains new archaeologies,
Rehearsing each indignity of pain
Behind the parched leaves glistening in the lane.

13

Behind the parched leaves glistening in the lane
Shade furs the cupboard rooms: heat piles a house
Of shapeless blocks where metal colours run.
Under a melting sickness, the obtuse
And ogreish language of a troubled sky
Flaps out its cloth of white and violet:
A speech where all the images elide
To pain of darkness and the pain of light
Till all the curtains rise, the whipped bough throws
A spray of ghosts across the hollow square:
A furious thickening in meshed foliages,
Blown droplets tingling as the flesh lies bare.
    The gathered waters drive for earth again,
    Diminished thunders breaking in new rain.

14

Diminished thunders breaking in new rain
And the rooms woken from a kind of dream,
The text and textures of their ground-work plain.
Calling each creature by its proper name
The conversations of a house unfold
One unremitting labour, which sustains
The open page, child's play and marigold,
The course the light sets in the hidden veins.
About the enclave of their pool, fish move
Across the shaken faces and the trees;
And through the air's refreshment all the leaves
Are congregating new-found silences.
    The chapter closes where the words begin:
    The summer opens where the days draw in.

15

The summer opens where the days draw in,
Leaves pressing home their small advantages:
The rim of summer, when great wars begin.
But there's a no-man's-land where skull-talk goes,
A hinterland, to breed new summers in.
The unfleshed dead, refusing to lie down—
What inch of sunlight gilds their vanishings,
Those penitential litanies of stone?
A silence runs beneath these silences,
And there our conversations must be held,
Where blood and stone proclaim their unities
And all the shadows cross on one high field.
    Behind the parched leaves glistening in the lane,
    Diminished thunders, breaking in new rain.

Effacements

The steady cedars levelling the shade
Bend in the waters of each diamond pane.
Furred cusp and sill: other effacements made
Where the armorial glass in bronze and grain
    Stiffens a lily on the clouded sun.
    The lozenge hatchments of the porch floor run

Far out to grass. The grave Palladians
Have gone to seed, long genealogies
Dissolve beneath their wet and gentry stones;
The leaves lie shaken from the family trees.
    As kneeling putti, children from the Hall
    Are playing marbles on the mildewed wall.

The letters and the memoirs knew His will;
All Spring contracted to the one hushed room.
White swelling grew, he passed the cup, lay still;
To the bed's foot she saw the dark spades come.
    They made such vanishings their deodands;
    The earth records what the earth understands.

Queen's May beneath his firstborn's coffin lid:
A lock of hair reserved, a brief prayer said.
'Each lifeless hand extended by his side
Clasped thy fresh blossoms when his bloom was fled.'
    Pain wrote in copperplate one epigraph,
    Then sealed the album's crimson cenotaph.

Behind locked doors, an audience of hymnals.
Tablet and effigy rehearse their lines
To cold light falling in a cold chancel.
Grammarians at their Roman slates decline
    Each proper noun. Bells close on a plain song.
    All speak here with an adult, eheu tongue.

But those frail mounds, brushed low by ancient rain,
Their markers undiscerned in the half-light . . .?
The last of day ghosts out a window; stains
The scarred porch wall, whose rough and honeyed weight
    Glows from the shade: stones so intangible
    A child might slip between them, bones and all.

Marginalia

1 *Leaf from a French Bible,* c.1270
Villeneuve-les-Avignons

A single leaf. Deuteronomy.
The untrimmed vellum keeps its pinholes;
Guide-lines rest uncancelled.

Under the Uncials' lantern capitals,
Whose gentler reds and blues are hung
On vines of gossamer.

The blocked ink scorches on the page,
Textus quadratus: bent feet hook
The linked chains of the Word.

A slant-cut nib works on; the skin
Takes texture. God is woven close —
The figure in the carpet.

And at the margin, crabbed, contracted,
Lies evidence the page was read. No more.
A commentary on silence.

Wind in the cloister's lost mandala;
Cold hands re-cut the blunted quill.
Somewhere, the child Dante

Sees Folco Portinari's quiet daughter,
And all the formal rose of heaven opens
On the one slender stem.

2 *Jeremy Taylor: The Rule and Exercise of Holy Dying, 1663*

Grey flowers in their pleated urns,
The frozen spray, the flying skull,
And the dark words' processional.

Thomas Langley Purcell. His book.
About the Doctor's text extends
A glossolalia of hands.

Where the Baroque moves into night
These light and active fingers thrive,
Cuffs frilled, the mood indicative.

And one small coffin, monogrammed,
Nibbed out with nail-heads. Nota Bene.
The period of human glory.

The sepia glitters and grows pale,
Memorial of a heart at school;
Praeceptor, student, work there still.

'You can go no whither but
You tread upon a dead man's bones.'
I read you still between the lines.

A shut book — as a church door closed
On congregations known, unknown,
On tireless monitors of stone.

3 *Rider's British Merlin, 1778*

The almanac's green vellum skin is rubbed
To duckpond water. There, stub fingers felt
For the cold patina of a broken clasp.

Under the stitching and the rough-cut ties
The 'Useful Verities fitting all Capacities
In the Islands of Great Britain's Monarchy'

Hold their rusticities of red and black,
Pragmatic as all northern frost and mud.
Look to your Sheep. Eat no gross Meats. Sow Pease.

The rightful owner still some revenant
From foundered England, scrawling his Receits,
Juggling his Honny, Daffy Elixir, Terpedine.

For the Rumatis, the Coff, For Nervious Complaints.
'Jesus Jesus Jesus My teeth doth ake
And Jesus said Take Up thy Cross and Follow me.'

No name. And yet by sleight of charm and simple,
In language stripped by dumb hedge-carpentry,
Some churchyard clay turned Goodfellow survives

The Bale-fires leaping out at Coalbrookdale,
The long-defeated children, all the sparrows
God seemed to have no eyes for when they fell.

Answers to Correspondents

Girls' Own, 1881

Queen of Trifles, we must consider your question
  Puerile and foolish in the last degree,
And May Bird, while we thank her for the letter,
  Must be less extravagant with the letter 't'.
Hester, imagine the sum required for investment
  If all could claim a pension who were born at sea.

Paquerelle, we fall back on the language of the Aesthetics:
  Your composition is quite too utterly too too;
Joanna, ask the cook. Gertrude, we are uncertain—
  What do you mean by 'will my writing do?'
Do what? Walk, talk or laugh? Maud, we believe and hope
  The liberties taken were not encouraged by you.

Constant Reader, if, as you hint, they are improper,
  We still do not see how to alter your cat's eyes;
Marinella, your efforts to remove tattoo-marks
  Are wasted. Wear longer sleeves. We do not advise
Cutting or burning. Smut, we could never think it
  A waste of time to make you better, and wise.

Tiny, the rage for old china is somewhat abated;
  Sunbeam, we are ignorant how you could impart
Bad habits to a goldfish. Inquisitive Mouse,
  Your spelling is a wretched example of the art.
Little Wych Hazel, we know no way but washing;
  Rusticus, may God's grace fill your heart.

Toujours Gai, your moulting canary needs a tonic;
  Xerxes, write poetry if you wish, but only read prose.
Cambridge Senior, we should not really have imagined
  It would require much penetration to disclose
That such answers as we supply have been elicited
  By genuine letters. You are impertinent, Rose.

Under the Words

Under the words, the word. There, for a moment,
Suffusions of slow light welled up and broke
Against unsettled surfaces.

Cadence against cadence. In their cut sheaf
Perdita's daffodils felt each torn spathe
Darken and cool as a deflected shaft
Broke head from stem. Wind set about
The resilient garden with a skirr of snow.
A globe of water on the sill inverted,
Subdued, then held these tiny languages:
Transposed scales of the Camera Obscura,
Aeolian harps, the unforced elements at play.

And now the white page, crowded to the margin—
So little space to write between the lines,
To place, against that weight and density,
The flying stance of wind or daffodil.
Under the footnotes, mesh and ply of tongues:
At windows, babel of obscuring snow.

Arena

The puppets danced. And the child stared,
Serious and still. 'How do they work, Giles?'
'By magic.' Then, corrective, the quick codicil:
'By string.' Fluent Magic, ominous, delightful,
Presents himself in terrors, expectations,
Or six Red Admirals on the buddleia.
String, though somewhat lacking in éclat,
Is equally at home here; serves his term
Among Kate's schoolbooks, helps to patch a quarrel,
Or by a reasonable beneficence allays
Those sudden sobs which shake a darkening bedroom.

Let their hostility advance our interests,
Like oil and vinegar. We must admire
Magic's deft web, String's workaday solutions:
This dance of Retiarius and Secutor
In our arena. But, should the crunch come,
May any Emperor on hand to give his verdict
Turn his thumbs down for String.

Twelfth Night

Our candles, lit, re-lit, have gone down now:
There were the dry twigs tipped with buds of fire,
But red and white have twisted into air,
The little shadow stills its to and fro.

We draw familiar faces from the wall
But all is part of a dismantling dark
Which works upon the heart that must not break,
Upon the carried thing that must not fall.

Needles are shivered from the golden bough.
Our leaves and paper nothings are decayed
And all amazements of the Phoenix breed
Are cupboarded in dust, dull row on row,

While branchworks set upon a whitened ground
Climbs out into a vortex of wild flame.
The substance of this deep Midwinter dream:
A scale of ash upon a frozen wind.

Our candles, lit, re-lit, have gone down now:
Only the tears, the veils, the hanging tree
Whose burning gauze thins out across the sky,
Whose brightness dies to image. And the snow.

Three Sisters

for Phoebus Car

1 *The Dry Tree*

Baron, as your fingers untie music,
Four seasons ebb and flow, the migrant cranes
Wind their slow skeins from climate into climate

And the dry tree shakes in its dance-measure:
Green arms, brown arms, linked in amity
Though bark strips from the cracked and polished bole.

A hover of dust: the patterned air discloses
Leaves shivered with light, gardens whose pages
Open or shut as dawn or dusk require.

But then the dissonance, the stopped movement.
The room withdraws to its four corners
And no God speaks a word of grace or power.

For now the windows thicken to a snow
Obscure and vast, with powers to impress
The old and unborn armies foundering there,

And a green Maenad dances out her frenzy
About the sapless branchwork of her fellow;
The soldiers break step on the silting road.

A dead bird, loosed from the retaining sky,
Stiffens: a twig of clouded foliage,
Eyes fathomless beneath their cooling hoods.

2 *Wild Grass*

In corners where a bird repeats one note
And clouds make blank faces over garden walls,
Hoisting into the blue and beyond the blue,

Time is a study in continuations.
But for you, Masha, should meaning cloud and fail,
Life, you say, becomes wild grass, wild grass.

In such corners, that is the wind's concern.
The bird repeats one note chack chack;
The wall dangles with sour and vivid cherries

Whose notes are discontinuous, abrupt.
The child at the keyboard picks, unpicks a tune,
Coding the air with lost ends and beginnings.

Here is the wild grass: a speechless city
Where threads of life are plied into a sampler.
Errands are run by the quick-dying creatures,

Each movement bright and urgent as the call-sign
Of child or bird tapping their counterpoint
Against the clock, the poppy-head's oblivion.

Listen to the soft thresh of their ground-bass,
Time and wild grass working, this way, that way,
Their tingling feather-heads, their cutting blades.

3 *The Curving Shore*

If only we knew. The garden fills with partings,
Rustling in summer silks and coloured streamers.
The green shell brims with music, and with pain.

A dance of insects, and the sun's gold beaten
Too thin for use: from all their cooling shrines
The little gods of place absent themselves

And broken paths return upon their traces.
A hopeless fork tricks out the garden seat;
The leaves are talking in dead languages.

Cloud upon cloud, the open skies re-forming:
Their echelons move off to the horizon,
Dipping westwards over the rim of the world.

Hand falls from hand; our eyes are looking down
Long avenues of falling trees. Apart,
The maple weighs its shadow out and down

On birds caught in a cage of air and grass
At work about their chartered liberties,
On pram-tracks lining out the sullen ground,

And music silvering the middle-distance
Whose bright airs tarnish as the bandsmen play.
The soldiers pass along the curving shore.

Horace Moule at Fordington

Distinction flowers on the darkened lawn;
    Sash windows open, the soft tassels shine
Where velvet rooms wait till the blinds are drawn
    On scents and silences. The heads incline
By their own gravitas; the tones exact
    Allegiance to a Church of England norm
Which has no need of this brief ritual act,
    This pause, outlasting all their hands perform.

Nobody smiles. They feel the sunlight there,
    Leaning across our shoulders; time unfurls
And we look back at these who look elsewhere.
    Their clericals absorb the light; two girls
Bleach out the summer from their swelling dress.
    A strength to set about the work of God,
This quiverfull of sons; quickened to bless
    Their strong maternal staff, paternal rod.

A neat, pale face, bare to the clearest bone,
    His eyes mere patches of uplifted shade,
The gifted one, the friend. And here alone
    All sight-lines meet, before the swerve and fade.
Our shadows will not press the grass apart,
    Or stir the cord on which their love is strung:
His finger still invites us to the heart,
    His throat uncut, the bastard son unhung.

Summer Palaces

How shall we build our summer palaces?
Will the girls bring us sherbet, and our gardens
Brown to the filigree of Chinese lanterns?

The Emperor speaks in a long robe of thunder,
Bruising us cloudily; his combs of rain
Dance out a dance of more than seven veils.

Islands of bird-music; storm-voices dwindle.
Across the blue, cirrus and alto-cirrus
Draw out an awning for our shade pavilion.

Swallows will sew our flying tents together.
We live as nomads, pitching idle camp
Under the white sheets blowing down the line,

Or swing on ropes to somersaults of grass.
The hasps creak upon our airy gallows,
Roofed by light, floored by a crush of earth.

The strong leaves curtain us; we know each scent,
The deep breaths taken behind swaying curtains.
We have become the citizens of green.

Our walls grow firm in fruit and knots of seed.
A dandelion clock rounds out the hour,
Blowing our time away in feathered segments.

The night lies warm upon a wall of shadows.
We lie as naked in our drifting beds
As the close moon, staining us with silver.

The Beach

And Langland told how heaven could not keep love;
It overflowed that room, took flesh, became
Light as a linden-leaf, sharp as a needle.

Today, the stone pavilion throws a window
Into the morning, that great strength of silver
Shawled from the climbing sun, and on four children
Alive to rippled beach and rippled water
Swaying their metalled lights in amity.

Hands build an airy house of meetings, partings,
Over a confluence of the elements
Here, where there is neither sex nor name,
Only the skirmishes of dark and bright,
Clear surfaces replenished and exchanged.

Black dancing in a hall of spacious mirrors,
Far voices, and the hush of sea on sand
Light as a linden-leaf, sharp as a needle.

The Chain

Above us, numbing all our dreams with tales
Of bad islands, infestations of gulls,
The metal broadsides of our great ark tower.
There hangs the chain: sealed rounds of iron
Whose chafe and rust, they say, ensure our freedom.
Paid out with its reel of swollen fathoms
Something of us gropes there, where deeper, deeper,
The hook works pain into a muddy craw.

Allow us good days, when, in certain lights,
Waves ease links to invisibility,
Gleam slips from gleam, all the lithe flexure
Polishes to a haze of blue and pearl.
So the stump dwarfs, gold-beating out Sif's hair,
Worked that obdurate element to a nature
Divine and animal: this was craft-work,
A clasp tendered by their slavish hands.

Our Pilot, arms akimbo, saturnine,
Rigged out in steeple hat and greasy frock-coat,
Stares back at us from his ancestral eyes.

The Gatehouse

Late. And though the house fills out with music,
This left hand takes me down a branching line
To the slow outskirts of a market town.
We are walking to the Gatehouse. Mr Curtis
Will call me Peäter in broad Lincolnshire;
Redcurrants glow, molten about the shade,
The cows are switched along a ragged lane.
Tonight, my son tousles away at Chopin
And a grandfather whom he never knew
Plays Brahms and Schumann at the same keyboard –
Schiedmayer and Soehne, Stuttgart –
The older, stronger hands ghosting a ground-bass
Out of a life whose texture still eludes me,
Yet both hold up their candles to the night.
The Gatehouse settles back into the trees,
Rich in its faded hens, its garden privy
Sweet with excrement and early summer.
New bread and sticky cake for tea. The needle
Dances across: Line Clear to Train on Line;
The lane is music too, it has no ending
But vanishes in shifting copse and woodsmoke.
The levels cross: a light and singing wind,
Arpeggios, a pause upon the air.
The gate is white and cold. I swing it to,
Then climb between the steady bars to watch
The station blurring out at the world's end.
The wagons beat their poésie du départ;
The lamps are wiped and lit. Then we, too, go.

The Spanish Train

The little Spanish train curls in the hand,
Its coat of many colours; the June garden
Will blow to seed, find new snows and sierras.
Somewhere, beyond the phlox, the cherry wall,
Goya etches: *tristes presentimientos,*
And though his rooms are hung with all misfortune
The train draws down a truthful patch of sunlight,
A radiance not yet underpinned by shade
Or lost in the earth-closets of the garden:
Neither the last mile to Huesca taken,
Nor the fixed siren set upon the Stuka.

Now, on a sofa, the child holds a word:
Spain, where the rain goes, and a wooden train
Quite serious in its unclouded paintwork,
Its yellow bright as any Star of David.
While the small fingers look inside a carriage
Or hook and eye the polished dolly-wagons
The flowers prepare their faces for the night.
Shutters and bolts are drawn. There are long journeys
Which must be made. *No saben el camino.*
There is no remedy. There is no time.
The little Spanish train curls in the hand.

from Notes from a War Diary

H.J.B. 1918-19

Madelon

This is the song the poilus sing,
  Madelon! Madelon! Madelon!
'Et chacun lui raconte une histoire,
  Une histoire à sa façon.'

A captured goat in the Canteen Car,
  Driven by 'Darky' Robinson,
Magpies, poppies and marguerites,
  Crosses are wreathed among the corn.

'Glorious weather all the time',
  (Shelling of Ablois St Martin),
To a walnut tree, by the Villa des Fleurs,
  Faux-Fresnay, Vaux, Courcemain.

He's never seen two prettier girls,
  With manners to match, pink dresses on,
Little Monique, Jacqueline,
  Mme Pinard — or is it Pinant?

And a stolen flight with Paul Scordel,
  (Back in time for parade at 9),
'Regardez à gauche pendant la spirale',
  (A mustard-yellow A.R. Type 1).

'La servante est jeune et gentille,
  Légère comme un papillon,
Comme son vin son oeuil pétille,
  Nous l'appelons La Madelon.'

But seven die at the Aerodrome,
  (Spads, appareil de chasse, monoplane),
'Forget your sweethearts, forget your wives',
  A Bréguet dives on a Voisin.

'Adieu Champagne, Villa des Fleurs',
To the Airmen and the waving corn,
The convoy leaves the little Square,
Beaumont, St Omer and Hesdin.

'Hope for a revoir bientôt',
Thirty-eight years, to the day, in June,
Back 'en passant' to the Villa des Fleurs,
To Faux-Fresnay and Courcemain.

Canterbury bells in the garden still,
The house closed up, the shutters on,
And Monique dead at twelve years old,
And the walnut tree cut down and gone.

What was the song the poilus sang?
Madelon! Madelon! Madelon!
'Et chacun lui reconte une histoire,
Une histoire à sa façon.'

Echoes

Like him, a survivor,
His notebook lays bare
The exact words cut
On the Picardy air,

And vanished Commanders
Are echoes, catcalls:
'Well, any answer to
Captain P's "Balls".'

A parade-ground whispers,
Under August sun,
Of deeds, decorations:
'There's more to be won.'

From a field in the field:
'Very gratifying to me.
I'm proud of you, and I want you
To be proud of me.'

That intolerable, urgent
  Intent of the dead,
Now light as a high-summer
  Thistledown head,

As his own unheard voice,
  Lips shaping 'C'est bon',
With the wind in his ears
  In that A.R. Type 1.

Fin

Madelon is left behind,
Her dispersal papers signed.

Spasmodic progress, mile by mile,
'Feeling pretty seedy still.'

Mons, Cambrai, Valenciennes,
By moonlight through Poix, Amiens.

Train-raid. Sleeping officers
Relieved of whisky and cigars.

'Khaki baked, new underclothes,'
Bread and bully, jam and cheese.

'Walk to Obelisk. Fine View.'
See 'Yes Uncle': a Revue.

March from the Delousing Camp
With books, flint implements and lamp –

'A trial for me.' Full marching order,
Kit-bag slung upon the shoulder,

Columns of four, and keeping step
To the harbour at Dieppe.

Channel-crossing. Dull, mild night.
South Foreland and North Foreland light.

Wait for tender. Disembark.
Fenchurch Street from Tilbury Dock.

Thetford under snow: the buzz
Of twenty different offices,

Buns and chocolate and tea,
'The Padre's curiosity.'

Cambridge station. Home at 10,
Lift on Royal Mail Van.

Soldier, scholar, he arrives
To what is proper: 'great surprise',

And ends the tale he never told:
'Wake naturally. Stiff bronchial cold.'

* * *

from CONSCRIPTIONS

National Service '52–'54

Scapegoat

Who are you there, huddled on your bed,
The sum and cipher of the awkward squad,
Your unbulled boots as round as innocence?
Nothing about you that will take a shine:
A soggy face, whose creases will not harden
To stiffening rubs of soap or these hot irons.
Unserviceable one, we circle round you,
Trimming our tackle to the Spartan code
And licking the parade-ground into shape.
Your fingers are all thumbs, your two left feet
Are mazy as your tangle of dull webbing.
Our silence covers up for you, our hands
Work your fatigues to parodies of order.
The small Lance-Corporal stares you cold and white;
His fingers itch to knock the silly smile
Off that dim face. You are our crying need,
The dregs and lees of our incompetence,
A dark offence, a blur on the sharp air.
These squares will not assimilate you now,
Whose discharge, something less than honourable,
Leaves our hutment stripped of all reminders
That something in the world could be in love
With pliant, suffering things that come and go.

Prisoner

You have fallen through it all: a cipher
Among the ranks you are reduced to,
The untamed residue of our glittering bull.
I watch your cropped skull bob between two Redcaps,
Those fresh-faced bullocks of our *Staatspolizei*
Stiff in their creases and their icy blanco.
We shall survive, but you are the survivor,
With nothing left to lose, or do, but time.
Your conduct prejudicial; crimed, impenitent,
Inured to having all the books thrown at you,
You do a Buster from the sliding train
Or slip your skin between the gated walls.
Your number's up in all the Standing Orders;
Our colours cannot call you. Your presence here
Is merely another absence without leave.

I link our wrists, trapping the bracelets on.
How shall we celebrate our Silver Wedding?
You flick a grin, strip with a knowing eye
My poor sleeves bare of sundry birdshit chevrons.
Oh, thing of darkness I acknowledge mine
By the conscription of a signature,
Your lightest word is ponderous to me,
But I will take you for the night's duration
While Moon ticks off the hours on her demob chart.
We shall walk out together, much in love,
For such a petty pace as you would creep to,
And hunt the shadows with a pack of lies,
Fusing the blue ghosts of our cigarettes
Till sleep takes you, and a sallow lamplight
Bones your neat head to cunning innocence.

Sentries

The inspection of shadows is our trade;
It is a business like any other.
Their stony faces are their misfortune:
Scratches of chalk on streaky blackboards.
When they speak, they do not raise their voices
And their complaints are blear and frivolous.
We know the fret and chafe of swinging iron,
The high wind slipping through a stretch of wire,
And little devils twirling in the dust
Abouts and roundabouts of wrinkled leaves.
Underneath the lamplight, the barrack-gate
Admits an air of cold authority,
Winking at our buttons, flickering ice
On swollen toecaps, lustrous ebonies.
We stamp their seal into a crush of gravel.
Let it all pass: ay, that's the eftest way.
We would rather sleep than talk;
We know what belongs to a watch.
Our relief is at hand; he will undertake it.
Wrapped in our winter pelt of pleated greatcoats
We fall into the consequence of dreams,
Careless who goes there, borrowing the night,
And will not stay to be identified.

* * *

Possessions

1 (*M.R. James*)

Shy connoisseur of ogee, icon, rune,
  Professor Prufrock ruminates decay:
Old Crowns and Kingdoms, food for thought and worms,
  All preservations held beneath the moon,
All sleeping dogs which can be kept at bay
  By gloss and glossary of well-thumbed terms.
His monograph extends by one dull page —
The last enchantment of his middle-age.

A fugitive from groves of Academe,
  The salt groynes running black from sea and sky
Betray his buried life down miles of sand.
  Low-tide: the beach is ebbing into dream,
Cold bents and marram-grass hang out to dry
  On swollen dunes. Deep in the hinterland
The calf-bound pages fox in white-sashed Halls
And stonecrop tightens on the churchyard walls.

Oh, blackthorn, unassuming country wear,
  Oh, modest rooms booked in some neat hotel:
Such pleasantries of the quotidian
  Are crossed by a miasma in the air.
Though bridge, or golf, might make him sleep as well
  And better than thy stroke, why swell'st thou then?
Bell, book and candle fail: the shadows mesh;
He counts the dissolutions of his flesh,

Appalled by faces creased from linen-fold.
  His textual errors thicken into sin,
Sour skin and fur uncoil about the night,
  Grotesques are hatching from the burnished gold
And countless demons, dancing on a pin,
  Repeat *a real fright, a real fright* . . .
Brandy, the first train home. The past dies hard
In stench and flare of bed-sheets down the yard.

2 (*Walter de la Mare*)

Grave summer child, lift your unnerving eyes:
Black bird-beads, drinking at the fluid light.
Whatever sifts among the unread leaves,
You pick your way, no Jenny Wren so neat,
Through Brobdingnag, here where the old aunts traipse.
Billows of shade, each face so loose and vague:
The stuff and nonsense clinging at their lips
As dusty, musty, as an antique wig.

A Conversazione. Old affairs
Flutter in broken heartbeats to decay,
Their ribbons fading into family ties.
Only huge, grown-up clouds for passers-by,
Cooling slow images in Georgian panes
Which close on rosewood, darken up the glaze
Of Dresden, and that chandelier which leans
Weak lustres over unplayed ivories.

The daylight shutters down, the white moths climb,
The house aches into whisper and desire.
You lie in bed, prim as the candle-flame
Pressing against the wing-beats in the air,
While nodding flower-heads and scents grow pale,
Mist curls where willows clarify, or dim,
As murmurs thicken in the Servants' Hall,
The callers gather; you are not at home.

3 (*Rudyard Kipling*)

The skull's base burning, and a swelling pain.
  A migraine of the soul: the Gods withdraw
Who made their Headings plain, and plain again,
  Whose word was Duty, and who spelled it Law.
    All slips beyond his reach—in the dark wood
    Where a wild nonsense claws substantial good.

The Lord was not there when the House was built,
  Though strength lay in the Mason's grip, the Sign,
And some who knew lie out in France, the silt
  And regimental headstones hold the line.
    Air thick with sighs; dead children stretch their hands
    To the Wise Woman, she who understands.

Is this the end, is this the promised end?
  Strange voices travel by Marconi's wire;
The crooked cells proliferate and blend,
  The cut runs counter to the hand's desire.
    A high wind wraps in winding-sheets of rain
    His House of Desolation, closed on pain.

They work about him at the razor's edge,
  Coax the indifferent stars to purblind sense:
Green simples, grounded by an English hedge,
  A surgeon, wiser than his instruments.
    He is returned, in ignorance, from the Gates.
    Under the draining tides the Day's Work waits.

La Primavera

In this, her kingdom, fictions of light gather,
Suffusing the pale flesh in a grave trance
Which is her thought, and whose desires are other;
The light airs hover over a ring dance.
The grace-notes hold there, and the hung fruits glow:
The riders all are fallen at San Romano.

The garden shaped my steps to her consent;
I read the text of her inviting hand,
Knew time to be dissolved, the censors absent.
No city shook its bell-notes on the land,
No New Jerusalem: mere trembled gauze
Whose veils disclosed and hid the garden's laws.

Then, when I turned, light lived from edge to edge
And each defining line became the source.
The glancing room conferred a privilege
To enter silence, follow out its course,
Turn to a window and exchange a stare
With frost laid cold on paths: untrodden, bare.

I knew this light, those breaths our dead resign,
That gift of tongues which holds authority,
Those dancers on some gold horizon-line,
That chosen ground. Her kingdom set me free
To share an hour unmeasured by the clock,
A space drawn freely between key and lock.

How, though, to judge and weigh the shift of planes
Or tell the climbing foreground from the sky?
Under the music come the dragging chains
And we are littered under Mercury.
Arrows must fledge the Saint; the pearl flesh dulls,
And Mary weeps there in the Field of Skulls.

La Derelitta bows her head, life thrown
To one eternal gesture of despair:
The flawless courses on unyielding stone
Pave the twinned elements of earth and air.
Savonarola burns; the scorched tears run,
Time throws a black smoke up against the sun.

Yet the globes hang there, and the star-flowers spill,
Speak mortal names on an immortal ground
Which is the ground of being, printless still;
The grove still sighs with music beyond sound.
Hortus Inclusus, and there is no stir
From haloed leaf and tree announcing her.

from TRANSFORMATION SCENES

Scott's Grotto, Ware

'From noon's fierce glare, perhaps, he pleased retires
Indulging musings which the place inspires.'
So, Scott of Amwell, fearful of the pox,
    Shifted his garden ground;
Quarried his Quaker frissons from the rocks
Till winking, blinking, Dr Johnson found
A Fairy Hall, no inspissated gloom.

But slag and pebble-bands decay and fall,
The glimmerings and lustres fly the wall
Whose rubble freezes to a dull cascade.
    The flint-work strikes no spark
And its begetter dwindles to a shade
Composing, decomposing in the dark,
Scribbling an Eclogue in a candled room,

Turning from Turnpikes or Parochial Poor
To the close circuits of this tilting floor,
Caves editing a Gentleman's Magazine
    Of cold and fading shells
Eased from their chambers of pacific green
To ride the beaches and the long sea-swells,
Then pressed into this ragged, alien womb

Which closes its dark backward and abyss,
Its miniature, yet Stygian emphasis
Upon your childhood voice, its terrors crossed
    Upon their own reply.
'Oh, if you're lost, then tell me *where* you're lost,
I'll kiss away the little deaths you die
And bring a light to raise you from the tomb.'

Ornamental Hermits

Veiled melancholy, mid-day accidie,
  Improvings of the landskip sheathed in rain:
We dwell, not live, shifting from knee to knee
  In robes as uncanonical as plain,
Nodding, like Homer, over knucklebones,
  (Cold vertebrae a starwork on the floor)
Serviced by birdsong: vespers, compline, nones—
  The cycles of unalterable law.
Our eremitic dirt, our nails prepare
  A mortification close to Godliness;
Under the tangled prayer-mats of our hair
  Each sours into his mimic wilderness.

The desert fathers, locked into each cell,
  Knew all the pleasures that retirements bring:
The succubus, the tumbling insect-hell,
  Assault of breast and thorax, thigh and wing.
Our bark-house windows, open to the sky,
  Show vellum faces, blind-stamped and antique:
A glaze of rigor mortis at the eye,
  A tear-drop crocodiled upon the cheek.
The unread Testament aslant each knee
  Will not dispel the simple moth and rust
Corrupting us, degree by slow degree,
  Into the scattered frictions of the dust.

We beck, we hoist our creaking arms, we nod,
  Blessing processionals of cane and fan,
Attuned to Paley and his tick-tock God
  Or swung in gimbals pre-Copernican.
And who dare call our observation vain?
  The celebration, not the celebrant
Will keep the ancient order clear and plain;
  We too are arks of His great covenant:
God's weathercocks, upon his axles pinned,
  The beads of an industrious rosary,
Tibetan prayer-wheels turning in the wind,
  Imprisoned things which float the bright dove free.

The dinky castles slip into the sea,
Though primped in paper flags and button shells;
A deckle-edge of ocean lips and swells,
Mining and slighting all our quality.
We see
Their looped and window'd raggedness decline
Into a sift of sand,
Hardened and barred as wind and Eastern wave design.

The scarlet toadstools pack themselves away,
The Christmas Cakes are crumbling to the plate;
The royal icings of their blood and state
Are one with all the snows of yesterday.
The tray
Still keeps the touch of sweetest marzipan,
Though little tongues of flame
Nibble the tissues into which their colours ran.

Gardens and gaieties are blown to seed,
The nymphs are crying down their party frocks;
The peacock straggles back into his box,
The secateurs are rusting in the weed.
Agreed,
The gnomons are in shade; each gnome's expression
Feeling the rot set in,
Is weathered slowly to a deepening depression,

But Homo Joculator starts again.
Though dyed in sinks of great iniquity
His mackintosh is black for luck, and he
Is whistling as he potters in the rain,
The pain
Absorbed by cupola and curlicue,
Those inching distances,
Castles in Spain which lend enchantments to the view.

* * *

Kingfisher

December took us where the idling water
Rose in a ghost of smoke, its banks hard-thatched
With blanching reeds, the sun in a far quarter.

Short days had struck a bitter chain together
In links of blue and white so closely matched
They made an equipoise we called the weather.

There, the first snowfall grew to carapace,
The pulse beneath it beating slow and blind,
And every kind of absence marked the face

On which we walked as if we were not lost,
As if there was a something there to find
Beneath a sleep of branches grey with frost.

We smiled, and spoke small words which had no hold
Upon the darkness we had carried there,
Our bents and winter dead-things, wisps of cold.

And then, from wastes of stub and nothing came
The Kingfisher, whose instancy laid bare
His proof that ice and sapphire conjure flame.

The Candles

A candle at your window, Reagan told us,
Light it for Poland. Forty years ago
A Polish airman taught me his 'I love you',
Breaking strange language out for me like bread;
His red and silver lynx badged my small palm.
Salt on a childish tongue: under the stair
Our fat soap candles lay; they lit the climb
To bed, and winding-sheets as cold as wax.
Their black wicks lifted to an oily stain,
Then sirens howled the moon, the chopper came;
The barrage lifted, and we still had heads.

Now, the Advent Calendar counts down
Shutters of light on swiftly cooling earth,
And what is known becomes a blue of wind,
A stretch of grass, dressed with the death of leaves.
The candles for our tree are boxed and still,
They nuzzle blind, delaying their soft fuses,
And though white blossom and the red of blood
Are twisted hard against our expectation
We shall not light them for a blue Virgin,
Or for a country where the luck runs out.
We press against the echo of a word

And think of English trees, an All Saints' Night,
The mists taking the conifers apart,
The mortuary chapel shut, but the gates open
For congress of the living and the dead.
There are the candles, glimpsed and guttering,
Behind the iron curtains and the wall:
Slight fires upon a shelf of Polish graves
Whose marble crisps to angels, crucifixions,
Names cut from the roots of another tongue.
The letters dance in black, and under them
Crimsons and whites of silken immortelles.

The celebrants, the host of lesser lights,
An autumn Pentecost, a gift of tongues,
Flutter themselves away in cells of glass
Against the cities: Wilna, Warsaw, Cracow.
Their slim flames, fuelled by the airs we share,
Still say 'I love you' in the grass and chill.
Here the sad suitcase and the paper bag
Are finally unpacked; our shadows lean
Across this pale of loosening flesh and bone
Which knew, that for a space, candles would flower,
Whatever comes with night to put them out.

A Borderland

This, then, is where the garden fronts the wood,
Where broken palings can be plucked like weeds
And thrown in brittle jigsaws on a fire
Which eats the heart out of good neighbourhood:
The nettles overgrow: the rusted wire
Follows the garden where the garden leads

Which is to nothing much: discourtesies
Of rags and mats which sop into decay,
A kind of murder where the clues lie down
In nests of earwigs and antipathies,
Dismemberings of green which stain the brown.
This is the place where children come to play,

To crouch their dens upon a borderland
From which the kitchen-door can just be seen,
A gravelled path, some garish flower-heads.
Here childhood and the trees make their last stand
About the half-bricks and the foundered sheds,
Play out their let's-pretends and might-have-beens.

Thick-set beyond Tom Tiddler's Ground, the wood
Is run by dogs—there it might start to snow,
Old Shaky-fingers pass his poisoned sweet:
The place where mother said you never should.
Its otherness might sweep them off their feet;
The wood itself has nowhere else to go.

A Midsummer Night's Dream

for Margaret

1 *Prologue*

One tree will make a wood, if you are small enough,
Watching with parted eye and double vision
Where Herne the Hunter, antlers caked with moss,
Rattles his withered chain, and quick as lightning
Knocks a green canopy to stump and crater.
You too can lodge there, rain and shadow falling,
Where the ants cross their tracks, puzzle your eyes
With the quick flicker of a coming migraine,
And, over pale saprophytes, the bracket fungi
Pull water out into brown cloths and cushions.
I crouched there once, upon that draughty landing,
Intent upon two ghosts who once were lovers
Rubbing fierce salts into their open wounds.
The darkness of the tree, the darkness of it,
Fostering such musty stuff in its own sickroom,
Twisting its accusations into mould . . .
'Keep promise, love.'
Here am I still,
Patient upon some tryst, some assignation
With those I cannot meet, but must pry out
At midnight, posted by a blasted oak—
The pencil licked and purple with love's wound,
The message scrawled in childish capitals—
Time bleached those lovers into grey, then silver,
Easing their passions into frond and tendril.
Now only witches nest here, cut from shiny paper,
Cloaks ample, capering legs as dry as broomsticks,
Their mumbled chins tight to their Roman noses,
Their cats in silhouette, fur jagged and standing.
Night passes round her boxes of black magic,
Crinkling a prelude to her puppet-theatre;
By the slight pricking of my thumbs I know
That things are coming to a head of branches
Whose windy rut rubs all the velvets down.

2 *Wind*

The trees push their loose heads about,
Dark hair angry with bee-voices,
The sob and growl of the unwelcome dead.
Fingers to fingers, feeling the world slide,
We call again 'Have you a message for us?'

Lost Uncles, prim Victorians
Frou-frouing in the dryish grass,
How easily our flesh parts for you,
Admits your sea-shell conversation
In the heart's echo-chambers.

These oracles are bare of priestcraft.
Only the deep-drawn sigh, the gusty breath
That there is no disputing.
We die into and across your voices,
The light fading, and ambivalent.

3 *The Elementals*

Bouts of quarter-staff, and the wind rising:
The soughing and the sighing start again
Under a clouded sky, an acid rain
Eating the heart out of the sluggish lake;
Over a ground whose flesh is gone, clean gone,
The trees must fail, their checkerboard is down;
The fairy-feller plies his masterstroke
And Delville Wood, High Wood, the Ardennes keep
The floored earth sour with an expended metal.
Carols and carollers have gone their rounds,
Church-doors are locked on prayer and profanation:
Wild-fire in stubble, swollen granary,
November windows garnished with false snow,
A ribboned maypole blunted into missile—
Is this arrested by a quilt of bluebells,
A stand of beeches blurred by your soft focus?
    'Do you amend it then, it lies in you.'

'And here we are, and wode within this wood,
Whose dark, unseasonable hearts congeal
To summer frost, sweat fire at winter suns,
Lovers whose tinsel chafes the tender skin,
Who know their nakedness, and are ashamed;
Our dance is broken, past our remedy.
Who will restore the changeling child again,
The coronets and cloths which keep a court
Of Lords and Ladies dressed in green and purple,
Fatten the pale hearts of the Shepherd's Purse?'
Unskilled in crisis-management, I wait
Under the Oak, whose Government is Jupiter,
The Vertues of whose fruit and bark will stay
The spilling of blood, and the bloody flux.
This I can tell, with sixpence in my shoe
And a sweet milk-tooth stolen from my pillow
When I was a boy (before these Civill Warres).

4 *Shadows*

Our shadows have disarmed us:
We rehearse their satyr-play,
Tumblers in the chalkish light
Which leaves a little of itself behind
Turning the corners of the leaves.

Such sweet and bitter fools
Caught in this rippled mirror,
Angels who dance on nimble pins
Where dry twigs fail to crackle,
And suffer the occlusions of the moon.

We are in love with our familiars
Who kiss with absent passion,
Undress our kindly flesh
And point, with close-webbed fingers
To grave and graver silence.

5 *The Court*

An old man's gums mumble for justice, justice,
Where trees decay to cloth and painted column,
Unhealed, unhealing: merest twists of gilt,
Alloys of tin and copper. There will be justice
Perched on a furred bench under apple-trees
At the conjunction of more fortunate planets.
Here there is bread and water: common, stale,
Wrapped in cliché, sickening to green.
Justice of walls. I too have known it,
Sent to my bed at noon, scaping the whipping
Until Our Father came, with a case of briefs,
His smoky chariot bolted to its groove,
Walking briskly under the sooted planes.
God must administer his own creation,
Confirm his delegates to their sober suits,
Fix their crisp eyes on fluid situations.
  'I never may believe these antique tales.'
Here, woods are stained, polished to artifice,
Bright in their inlay; in their deaths they curl
Cold, unresponsive to the play of flesh
Which works its passions out against the grain,
Intent on cupboard-love and closet-secret,
Confabulations of the cabinet.
The floorboards tick as Love's old Counsellors
Trip down wrong passages, tap bedroom-doors
Behind which skirts are lifted, stockings bare
Their dexterous enchantments, their bend-sinister.
At stud the bitches couple and uncouple:
Castlemaine, Ariadne, Lily, Aegle . . .
Out in the fields, beyond the swerving headlights,
The bears are bushes and the bushes bears
Where the white girls are sobbing in the wind.
Under the aegis of oak, ash and thorn
Even these halls must suffer consecration.

6 *The Green World*

The green world, figuring it out,
  Renewed, Arcadian,
How could these couples ever doubt
In snowfall, hail or waterspout
  That Spring would come again?

For Aucassin, for Nicolette,
  For Darby and for Joan:
The sunlight sharpening the wet,
The Gamut, Octave, Alphabet,
  The softening of stone.

Obedient to natural law
  The folded lovers lie,
Their senses quickened to explore
A Masque of Brightness played before
  Conspiring hand and eye:

The cuckoo's double-fluted trick,
  The lanterns in the grass,
The colours of May's rhetoric,
And, fire from Winter's musty wick,
  A chestnut Candlemas

For Flora, ribboning the scene,
  Enskied in white and blue,
Who reconciles each might-have-been
With was, to come, and in-between
  To make it new, and true.

7 The Lovers

'Whose voice is this, shouting the wet woods down,
At home in a goatish sabbath of black smells?
The flailing branches in their russet suits
Know that our tongues are tripped, and all they meant
Runs idle messages into sodden ground,
For all the lines of briar and vine are crossed
On which our words were strung, by which our fingers
Trailed their unlucky scrapes into this no-place.
The ground, puck-hairy, coarse with violence,
Whips out its barbs to suck hot beads of blood
From blue love-bites where snapped thorns rankle.
By the slung moon's pale ignis fatuus,
Mounted upon some bucking lubber-fiend,
We are led all night a byway, hunting down
Forgotten courtesies of the drawing-room,
Triangles of dark hair between pale thighs.'
        'Follow me then
                                        to plainer ground.'
'The Puck is busy in these Oakes, rough Robin
With all his saplings sharp in Lincoln Green,
Their bows drawn at our ventures, clouded now
Into old scarecrows and a creaking darkness.
We will turn out our tattered coats upon him,
Loosed from that School of Night in which we studied
The shameless quarters of our hearts and loins.
Now, strewn with strawberry leaves, we hold our flesh
Against its resurrection: careless limbs
Stroked upon limbs, the breathing easy
And our own voices dreaming us to sleep,
Spelling us out more clearly than we knew.
Above our heads, kind stars will steady us.'
I watched their waking from and waking to
On a bank sweet with natural grasses, dews,
And a pied dawn to bless their mortal houses.

8 *Constellations*

Out of their fixed sphere now:
Free-fall of beast, girl, hero —
Lost in O Altitudo,
A space beyond all spaces
To take our breath away.

As a child figures it out,
Tracing the master-plan;
Joins up the dots again,
New bearings taken
On signs, houses, occultations.

Throned, garlanded in fire,
They lean their absences
Over our old gods, guisers
Whose speech becomes ellipsis.
Each to his chosen star.

9 *The Mechanicals*

The problem is, how to disfigure moonlight
And keep the accidentals of the moon,
The onceness and the hand-in-hand of it.
The green plot thickens, and her clouded globe
Bears witness to these rings and roundelays.
Ovid is banished where the Black Sea moans
Tristia, tristia — still the mulberries
Fatten their milky drops on Thisbe's blood:
Lovers must die, and garden walls must crumble.
I knew you, Peter Quince. One careless summer
I, tender juvenal, watched your hands at work,
Wood in sweet flakes curling in whitish dust.
The bat you made me sent your daisy-cutters
Out to the boundaries of childhood.
Old friends, the wood was a fine stand of timber,
The Carter's bells have jingled it away.

'So please your Grace, the Prologue is addressed.'
Only the barest boards, the rawest deal —
But as you spread about your cloth of gold
The tuppence coloured and the penny plain
Make harvest landscape, all its fecund light
Fingered by Midas with his ass's ears.
I have no fears for you as you troop in.
I, too, wore glad-rags, held the wooden sword,
Dubbed carpet-knight by a grave audience.
Something can be made of any lines
By those who, tinkering in draughty sheds,
Patch up the broken to the whole again.
It would be good to leave a window open
When bushes, dogs and lanterns go to sleep
And the late brands are wasting into ash.
The great oak bressummer, the hearth's roof-tree,
Will hold enough of moonlight to get by,
Struck from the adze-marks of your shaping fingers.

10 *Moon*

It is radiance on its moving staircase
Parting the drapes of always night,
The tide slipping her floodgates,
A gravity which draws us out
To clear silver beyond our pale.

Expert with wound and bow
She runs her circles round us;
Littered under this mercury
We must accept her ground-rules.
Our glances travel at the speed of light.

We offer her no histories,
No white slow-motion surgeons
To graze dry pastures.
Rather, the simple love she mirrors:
For ever new, for ever at the full.

11 *Epilogue*

Observe the rites of May, betake yourselves
To wakes, summerings and rush-bearings,
Wash with hawthorn-dew on a bright morning
For beauty's preservation; hugger-mugger,
Enveloped with a mist of wandering,
Lose your virginities in wayside ditches.
Your lavish Queen, made bold in this Floralia,
Sits in her arbour by the painted pole,
Under whose standard Justice is agreed.
Dance for Jack-in-the-Green, your high roof-boss,
The oak-leaves foaming from his wooden mouth,
Before your bonfires, garlands, junketings
Are levelled by a thundering ordinance,
Your landscape raked by new theologies,
Your hedges rip-sawn and the wood stubbed out,
Dragons' teeth sown in the poulticed gums.
  'Think you have but slumbered here
  While these visions did appear.'
I walked in the astonishing light of trees,
A tenant only of their close estate,
Where the Great Oak, scored and furrowed,
Grappled his roots through subsoil into clay;
His branches, solid in the rising mist,
Have housed us all, and burned our oldest dead.
The Wodewose, with matted hair and beard,
Who knew the springs of love, walked there beside me,
As dark as any heart. Night fell again;
I took the key to her dark wardrobes,
Fingered the sloughed skins, the heap of cast-offs.
The spars fell loose; their sails were nets whose mesh
Caught nothing more than a little dust and air.
Now, out of my text and out of pretext
I pray for amity and restoration,
Who have dreamed waking-thoughts, woken to dreams.

* * *

Green

The garden laced green stuff about the two of them,
  Soughing, shaking its points of wet light;
Under and over, leaf upon quick leaf,
  The seasons following, folding them out of sight.

And the web shook: quietly, sometimes fiercely,
  The branches high-strung over long grass;
The back-gate was choked by the Zéphirine Drouhin,
  It became easy to sit and watch days pass

Leaching colour from the deck-chair canvas.
  He was unmanned by what had vanished; she,
Under the spread and weight of the blue sky,
  Set out the bright things for a late tea.

One year goldcrests came; another summer
  Herons took two fish from the cloudy pond.
The gate, the green, the cool rooms had decided
  There was no future in looking much beyond

Bees in the hyssop, a wild honeysuckle
  Cupping its warm crowns against the wall,
The mild whispers of things growing, being, dying,
  Fingers of light slipping into the front hall

Where the house, chastened by wax and cut flowers,
  Split all the green to something without a name,
The sun and the moon stretched out their warm and pale
  On the spare bed to which no visitor came.

A House of Geraniums

It is calm work, remembering their names
In a house of geraniums and white stucco,
Looking at death and looking at the sea,
Salt on the wind filming its weather eye.

Each photograph is purest ectoplasm,
Part of the way lights crease the window-sill,
A teacup's ring is glimmered on the table,
Kind thoughts lie pencilled in a Fairy Book.

But all these wishes, all these visitations?
'I know she was with me on the day she died.'
Sheet-lightning: air stuck in sultry vapours
That she dispersed with a particular joy.

And other, simpler wraiths who keep their patience
As we arrange a mystery about them,
Who, if we find them voices, only say:
'We were; you are. Why should you ask for more?'

The dead are anything that sighs in millions:
The tossing flower-heads of Queen Anne's Lace,
Moths trawling the dark, those waifs of snow
Flocked against glass, formations of the night

Which grows towards us: ghosts talking of ghosts,
Compounded of old walls, old bones, old stories,
Watching an inch of sun slip to the West,
Playing the revenant to this house and garden

Sleepy with cats down a remembered lane
Where unaccustomed eyes look cleanly through us;
Pity our grey hair, unfamiliar pauses,
Our tongues which trip so lightly over our graves.

The Key

Neither leaving, nor wishing that they were staying,
        They stood at the gate,
Cinnabars on the ragwort, the wind crossing
        Swords in the bents, the light
        Moving about.

There was nothing else much that needed doing,
        The key under the stone,
The stone under the grass, the grass blowing
        Under time gone,
        The fences down.

And already the rooms were changing, the ochre curtains
        Fading to blue,
The carpet figuring out a forgotten pattern,
        Leaves learning to grow
        As they used to.

Cows grazing their shadows, a clock ticking,
        A pause to share
The hour with anything else that needed saying.
        A note lay under the door
        To be read last year.

They stood there still, as if time, the grass tossing,
        The white stone,
The key to the blistered door that was always missing
        Could ask the two of them in,
        Or wish them gone.

Genio Loci

Brick and mortar for the ghost
Slipped between our first and last,
That beneficence which came
When man called his hearth a home;
From a roaming wilderness
Cut the mystery of place.

Overhead the star-beasts pass
This loose gazebo, winter-house,
Patched with glass and cockle-shell,
Cobbled up with flint and tile,
Where the garden guardian
Hosts his feather, fur and fin,

And, as coloured lights go down,
Glazing rime upon the green,
As the winds blow colder through
Spaces leaded bough by bough,
Keeps at this dark terminus
The empty nest, the chrysalis.

The Gardens

You cut the garden where a garden grew,
And surely as the calendar made days
Appear and disappear, his borders, beds,
Became a waste of ashes, wind and weed.
The little sherds and prints of china knew
That he and others threaded out this maze,
And long-lost flowers puzzled out their heads
Before the silks and satins blew to seed.

Your garden cut into those gardens, where
Old spade-work went to ground: a saddish blur
Of brick and half-brick, fingerings of green
Which made their lack of substance all too plain.
You shook the memory out in tides of air
With such Spring-cleaning, such a leafy stir
Of blues and whites, such branches tossed between
The sky's conspiracies of sun and rain.

But country ghosts will talk of country matters
And show their skills in sign and countersign:
Their broken pipe-stems chalking up the clay
Through which rust nuggets of cold iron climb.
In any song a thrush or blackbird chatters
Of what is his, or hers, or yours, or mine,
You'll find the resurrection and the May
Of those who had their world, as in their time.

Ragtime

He moves about his traps of light,
This errant and unerring thing,
To catch the dust with double paws
Or damsel flies upon the wing,
  And certain of his small terrain
  Will cut and run and twist again

To take, as due, bread from the hand,
Or flash his coat of cinnamon
Across cut grass and sun-stroked stone,
His come and go, his came and gone
  Exacting from the summer air
  A sense of being here, and there;

Or sleeps, as if that sleep was all
That sleep could be: a loose repose
Where all the garden's coloured quilts
Conspire about him to disclose
  Their buzz and bloom, confusion clear
  To one inconstant pricking ear.

In all these rituals and rounds
Each sinew, nerve and sense agrees
That his mysterious web of life
Should traffic in simplicities,
  And offer belly, throat and paw
  As hostages to human law,

As if such trust were possible,
As if his amber, secret eyes
Held nothing in their narrow glass
Of revelation, or disguise:
  Clay woven into voice and purr,
  Caressing hand and molten fur.

Cat's Cradle

for Tim

A cradle made of string
  For hands to ply and plait,
Teasing a see-saw thread
  For one old journeying cat.

We swung him down the night,
  Mewed in his wicker ark;
His dancing eyes lit up
  The soundless dews and dark.

Summers of goose-grass led
  Down paths we had not made
To nests of tabby gloom,
  Soft play of sun and shade.

Long absence held our breath;
  The house grew wide and bare
Until he called us home.
  I cut out of cold air

His face with its sharp look,
  His fur grown stray and dim—
For these long miles of sleep,
  Let the earth cradle him.

from Erotion

Martial V, 34

Father and mother, now
I trust this child to you —
Erotion: all my thought,
My kisses, huge delight,
Who lived but six days less
Than six cold solstices.
Fronto, and you, Flacella,
Take your best care of her,
So small, and out so late,
Lost, scared of the night
And slavering Cerberus
Gaping his monstrous jaws.
Help her to play on,
My name brightening her tongue,
Frisking beneath the heads
Of those old greybeard shades —
And earth, your lightest turf
Will be enough
For one whose flying feet
Carried so little weight.

Martial V, 37

For lost Erotion
Heap on comparison:
A voice to mute the swan.

Skin softer than fine wool,
Purer than whorled shell,
Snow, ivory, lily, pearl,

Hair a more golden fleece
Than any girl's cut tress,
The quick fur of a mouse,

And breath a sweeter balm
Than rose or honeycomb,
Amber, warm in the palm.

The peacock shone the less,
The squirrel fell from grace,
The phoenix — commonplace.

But Fate is bitter, sure;
All that is left of her
Lies ash on a fresh pyre

And Paetus mocks my grief:
'My wife dies, yet I live —
And you cry for a slave!'

What courage, my old friend,
Sticking it to the end,
Blood-money in your hand!

Martial X, 61

Here lies a childhood lost,
Quickened into a ghost.

Six winters brought their snows:
Fate spun the bobbin loose.

Stranger, you who inherit
My small protectorate,

Bring, to confirm your reign
Gifts to her marker-stone.

Honour your Household Gods,
Respect the much-loved dead.

Be happy. No dark years,
No further stones. No tears

To stain your patch of sun.
Think of Erotion.

Under the Barrage

Schlafe, mein Kind
In your mother's bed
Under the barrage.
The soon to be dead

Will pass you over;
It's not your turn.
The sheets are warm
But they wil not burn.

In your half-dream
Sirens will sing
Lullay, lullay,
Lullay my liking.

A saucer of water
For candlestick:
The night-light steadies
A crumpled wick

In a house of cards
In a ring of flame;
The wind is addressed
With a different name.

Schlafe, mein Kind
In your mother's bed.
Under the barrage
The soon to be dead

Lie as you lie,
And will lie on
When the dream, the flame
And the night are gone.

The Other Side of the Hill

Who is it mooning about in his half-lit room,
Sharing my name, the set of my bones: a boy
Hung in my chains of words, puzzling it out
Under suspended judgements, gagged and blind
While Time poisons him cloudily under the door.
What are these thoughts I badger him into thinking,
These possible routes I map for him, joining the dots,
Holding his secret writing up to my fire?
Oh, he is poor Jim Jay, stuck fast in Yesterday,
With his party tricks, the clumsy deck of snapshots
He shuffles, cuts and deals, those blind fingers
For what I have dropped and lost on his bedroom floor.
I would like him to put our heads together, confer
About old hedges, pavements, the dazzling film
Which is racing out of sync with its dusty soundtrack.
Perhaps, though, all he can say is 'I was there',
As if 'there' were a special place, or a right of way,
Not a sentence shared with the other prisoners:
The sad Italians behind the bulldog gate
In the flat, unhurried fields, who turn and wave,
Ridiculous in their Commedia diamonds, moons,
Our next-door neighbour, jabbering lost Polish,
The bitter stench of her house dark in our nostrils.
She paces for ever behind her shining fence-wires
Which sing as I pluck them: the Chekhovian sound
Of a nerve thrilling under a huge dull wound.
Absorbed, at home in this peculiar country,
He leans from my Tower of Babel, anxious to help me,
Then settles back again into his thousands,
Swinging his satchel, counting his souvenirs,
Reading the signs on the other side of the hill.
For old time's sake I offer him this future:
Another chance to make things up. After all,
I am the victim in his puzzle-picture;
He is the child that I go looking for.

Diet

They gave me bread and brains and buttered porridge,
A dusty glass of water, stale and sweet,
To take away the sickness of the roses
That climbed the bedroom walls and could repeat
Nothing but roses, roses, till their nonsense
Crept sweating round me in a winding sheet.

They pushed their gravy trains into my tunnel
And stuffed me up with garden slugs and snails.
My salad days were crisp and limp as lettuce,
Then all the clockwork toppled off the rails,
The little foxes jumped from all the foxholes
And ran about the world with blazing tails.

They rubbed their hands in sorrow and in anger,
Then put the crumbs out on the window-ledge:
The crows preferred to gobble up the eyeballs;
And all the children's teeth were set on edge.
The devil spat on all the plumpest brambles
And sharpened up his nails on every hedge.

They laid the tables with their best behaviour,
And stuffed my manners in a napkin-ring.
I told their boasting turkey it was Christmas
And sang that I was happy as a king.
Then when it lay in bits of skin and rubble,
I cracked its wishbone, wished like anything.

They never thought, when wishes turned to horses,
And all the beggars rode away, a set
Of crows and foxes, eyes as big as saucers,
Would fight for any sweetmeats they could get.
I egg them on by eating to remember;
I buy them off by drinking to forget.

Bigness on the Side of Good

I slip down alleys to our other house,
Where Granny Puff jollies us into laughter
And a million snails live in the wet ditches.
I swing on walls there, crouch in a Ford's skeleton
Bunkered by long grass, while all about me
Striped cinnabars are munching up the ragwort.
My Uncle keeps a monstrous pig, an airship,
As pink as Churchill woofling in his den,
All chubby-chopped, rooting for Victory.
My Uncle Tommy is a man you must look up to —
A Captain Flint, the proper size for uncles.
His broad back can take a deck of cousins
Giggling and swaying through a summer day;
Churchill gives his famous vulgar V,
Swelling his voice into a buttery grumble.
I take his name apart: the rich grey church
Set heavily upon its smooth green hill.
My Uncle's pig smells rich and rotten-ripe
As I drape its backside with a Union Jack
While cousins fall about with lunacy,
Shelved on the wall like Beatrix Potter's kittens.
Later, we find the flag gone very missing,
Marked, gorged upon, and inwardly digested;
Our stomachs have to swell with pride or tighten.
The pig died, surfeited by patriotism:
John Bull and the pig, wrapped in the flag or round it.
My Grandad is reciting 'Barbara Frietchie'.
'Shoot, if you must, this old grey head,
But spare your country's flag!' she said.
A pig, of course, would not care to respect it.
My Uncle is the most famous man in Rasen;
He does not know that I have killed his pig.
Everyone greets him as we walk together
To the Observer Post out on the Racecourse
Where German bombers in black hang from the ceiling.
Having fought right through the first World War
He'd know the right way up to run a flag,
How to cure hams, how to bring home the bacon.

The Old Frighteners

When the impossible grottoes hunch their backs
Under the crinkled stars, we dream to find
The sirens blowing down the bedroom chimney,
Far, far away, and paler than the wind

On darkness hanging from the window-panes.
A pail of water and a pail of sand
Will keep the devil in, and devils out.
What is it that goes round the house, and round,

Dragging its tarnished ribbon off the spool,
Creasing its wicked face upon the moon?
Unsteady things call to unsteady things;
All the slack fur is rubbed against the grain.

Whoo-hoo, whoo-hoo go the old frighteners,
Their clammy sheets twisting between our toes,
Banging the doors about in the grey ghost-house,
Pushing their long dull fingers into eyes

Sticky with sleep, half-opening to find
The sad lights fluttering their yellow-brown
Off the soft edges of the bannisters.
We float downstairs upon the ache and drain,

The unison, the dying aah of things
Traipsing their nonsense back into the grave.
Then the bristle of night standing on end:
The guns chuckling out in the witch's grove.

Double

She says he came back that night. She lay awake,
The usual skeletons propped up in their cupboards
With best-china faces. The guns began to bark;
The shrapnel pattered. When more stones fall than usual,
Consult the Sibylline Books. I do consult them,
But the pale snapshots in their padded albums
Are ancient lights, ambiguous oracles.
We know, though, he was out, and at his post:
One of those Augurs in their stencilled helmets
Reading the scrawled sky, the city's entrails.
Stoics and bombs whistle as lightning strikes;
The skeletons crooned, bouncing up at keyholes.
She, hoping, praying that we would make old bones,
And, as she swears, with all her wits about her,
Turned to find him there in the smallest hours,
Moving about the room he had shored up
With baulks of timber. Under those auspices
They chatted together for a little while,
Her fear draining away, the heartbeats gentle,
Before he closed the door upon himself,
Blacked out his shadow on that darker shadow
So tightly they grew indivisible.

Later, much later, she got out of bed
And laid a breakfast of crossed purposes,
While he, rubbing the fires from his eyes,
Wrote out a testament, alas, gone missing,
To clear himself of presence without leave
That night when Love and Duty had his name
So clearly down upon their distant rosters.
Martial says 'Dream of yourself, or stay awake.'
She says that he came back; he still denies it.
I was asleep. What I saw I will not tell you.

Going Out: Lancasters, 1944

'They're going out', she said.
  Together we watch them go,
The dark crossed on the dusk,
  The slow slide overhead

And the garden growing cold,
  Flowers bent into grey,
The fields of earth and sky
  Losing their strength to hold

The common lights of day
  Which warm our faces still.
Feet in the rustling grass
  We watch them pass away,

The heavy web of sound
  Catching at her throat.
We stand there hand-in-hand,
  Our steady, shifting ground

Spreading itself to sand,
  The crisp and shining sea.
Wave upon wave they go,
  And we stand hand-in-hand,

The slow slide overhead,
  Stitched on a roll of air,
As if they knew the way,
  As if they were not dead.

The Wooden-heads

I am learning nothing at incredible speed,
Weltering in ink: a blotted copy-book,
My nibs and fingers crossed unhopefully.
Crumbled from the soft pages of a library book,
The wooden-heads are punishing my daydreams.
Zombie-skittles, they run rings round you;
The circle closed, the victim vanishes.
Having cleared London, they will start on Cambridge.
Ropes and wall-bars offer no escape-routes;
Matched at bantam-weight, I bob and weave,
My gloved hands pummelling the air.
They have caught Hodges; by the cycle racks
I saw him, red-faced, simple, yellow-haired:
Now empty, strangled in his father's toolshed.
The foul-mouthed kapos of the middle school
Are doing a roaring trade in small extinctions,
Trawling a dragnet for us small-fry.
The game is gas chambers. Into the labs with you,
Heads pressed hard against the bunsen burners
Whispering you to sleep. We lurk in crevices,
Closed by dark wood, secrets, grey flannel,
A master with one eye like a scorched Cyclops,
And feel our little strategies go down,
Slipping through all the circles of the underworld.
The gowns and blazers form their broken ring;
There is a gap still I can wriggle through
If I can only purge away my scraps
Of dog-Latin, wrong notes, false equations.
Best to go to ground in the reeking bogs:
Unbolted stalls, chads, pig-troughs of urine.
The wooden-heads are out looking for minds to bend,
My satchel of trash blocked up in their beast-world.
I see Hodges swinging behind a door,
His childhood flushed away, his bright blue eyes
Still laughing at me over his knotted throat.

His Face

His face is everywhere: a slab of paste,
Dabbed forelock, hot pale eyes. His lair
Is a wild nest of burnt metal, shadows,
But I know a good-luck works between my fingers
In Granny's brooch, the soft red-leather Kipling
Where his black-magic cross stands gold and upright
Or swings its level arms over its shoulders.
Its firework cartwheels on his wooden sleeve
Into the pages of a million schoolbooks
As he strikes air rigid with his quivering fingers,
Brooding impossible rages with a tongue
That skids unbraking over the crackled ether.
His flat hand launches out his throbbing arrows,
Flung through classroom dust and the long nights
When scuttle helmets bob in rows before him,
The banners dip, the oily torches flare,
The high goose-steppers kick the air aside.
Tiring, he twists his fingers in his mouth,
Blows out his brains, dowses himself in petrol,
Blazes himself away, runs wild with Eva, Blondi,
All the cold wolves who stuck to him like glue
And slavered to pull down the sun and moon.
Now, he shrinks to a well-practised doodle,
A dead thing which thrills my poking fingers.
When I peer at the cracks between the clouds
I know his fleets will rise to his command,
Their crushed bones hauled and straightened from the tomb,
The cowlicked skulls perched in their greenhouse cabins
Blinding their way across the sulky ocean.
In the School Dining Hall I take my Spam
And cut it to a neat, pink swastika.
I eat it slowly; it is his communion.

V.E. Day

Carpamus dulcia: nostrum est
Quod vivis: cinis, et manes, et fabula fies.
PERSIUS: Sat. V

Noticing oddly how flags had been rubbed thin,
Bleaching in shut drawers, now unrolled
In blues, reds, their creases of old skin
Tacked on brown lances, headed with soft gold.
    Clothes-lines of bunting,

And light fresh at the front door, May
Switching the sky with stray bits of green,
The road levelling off; the day much like a day
Others could be, and others might have been.
    A woman laughing,

Sewing threadbare cotton to windy air,
The house open: hands, curtains leaning out
To the same gravel, the same anywhere, everywhere.
Birds remain birds, cats cats, messing about
    In the back garden.

And a table-land of toys to be put away,
To wither and shrivel back to Homeric names.
Scraps gathering myth and rust, the special day
Moving to its special close: columnar flames
    Down to a village bonfire

In which things seasoned and unseasoned burn
Through their black storeys, and the mild night
Fuels the same fires with the same unconcern:
Dresden, Ilium, London: the witch-light
    Bright on a ring of children.

Night, and the huge bombers lying cold to touch,
The bomb-bays empty under the perspex skull.
The pyres chill, that ate so fiercely, and so much,
The flags out heavily: the stripes charcoal, dull.
    Ashes, ghosts, fables.

Jacob's Ladder

When searching cones of light
  Transfix you at your station,
And the radiance foretells
  How night will claw you down
Into a soundless glow,

Who do you see on Jacob's ladder
  When the rungs are blazing,
The children rise in ash,
  Sifting the dark with faces?
Non Angli, sed Angeli,

Pitched from seas of flame
  Into such absences,
Tossed as Dante's lovers
  In cyclones of desire:
Brands plucked from the burning

Quenched in a great cold.
  For flesh packed in cellars,
Bonded, swollen, molten,
  Shrunken out of substance,
Whose hands make cradles?

When the air is full of names
  Blowing in the slipstream,
The city lies in cinders,
  The moon, immaculate,
Crooks her sharpest horn,

Who do you pass on Jacob's ladder
  When the rungs are blazing?
What traffic have the dead,
  Tongues burned, eyes blinded,
As the squadrons turn for home?

Blackout

Surely the light from the house must stay at the glass
And pull itself back, into the room again?
For the light is ribbons, faces; it cannot pass
Out into the loose garden, and the rain
Which scrats at the pane with absent, occasional fingers.

And all our drapes, curtains, tackings of cloth
Closed on these glances, glimmers, openings,
Where the cold glass sticks tight to the hung moth
Quivering its white thorax and plumed wings —
How brief they are, how pale those lookers-in —

Are teaching the light it has nowhere else to go
But round the angles of things, filling the square
Where a sofa thickens its back, the bookshelf row
Scatters in flakes of gold, the arm of a chair
Wobbles where shine slinks off, and into the corner.

I slip round the door and the years, into the dark,
My feet risking the gravel, the wind and wet
Slight on my face, but the house is shut: an ark
Where no parting or pinhole eases to let
My eyes build back the room where I am sitting,

Curled on a cushion, dragging the hours to bed,
As I stand outside in the heaviness of the night,
The child in the chair in the room turning his head
To the out that is bonded out from his cage of light.
I am hunting still for a way to return his glances,

For the house is a black cloud stopping the lawn
From sheering off into further wastes of grey;
But the child inside knows well that the blinds are drawn
Not to keep the light in, but the dark away —
And the future, its blind head nuzzled against the window.

Service

Hearing the organ stray,
Wambling slow time away
In sit and stand,
A candle-bracket cold to hand,

Watching the shadows pass
Under this burning-glass:
One bright eye
Of lapis lazuli

Over an old saint's head,
A running maze of lead,
Robes of blood,
And little understood

But the high lancet's blue
Which makes the whole tale true
And is definite,
Shifting its weight of light

As afternoon lies dying,
The trebles following,
Stone gone dim,
God down to his last hymn.

Magic Lantern

The magic lantern show is nearly over.
The bilious frog in his tasselled smoking-cap
Croaks and goggles fubsily in his box;
Ageing monkeys, faced in gold and scarlet,
Twirl themselves off with their barrel-organs.
Goodnight to monocles and waxed mustachios,
Priggers and prancers of the Victorian order.
Light now draws perfect family circles,
Grey as smoke-wreaths, dim as muddled homework:
My mother, profound and sleepy as a doll,
Grandad, young and bowlered, slipping himself a grin,
Granny laughing, a dog — wild shreds of hair,
Long nose, black eyes come from the dust
To see nothing that once saw something.
Now for the tailpiece: Grandad's fingers
Twitch past a small black train, mad as a spider,
Tiptoe across a bridge across a ravine
Across green tumbling across to a blue river.
The train pulls gruffly out of Lincoln Central;
My eyes choke on coal dust, search the wind
And clanking signals for some huge delight
Cloudy as last year's pain, bright as a dewdrop.
There is a handle, polished as a nut,
The key to the last slide lifted from the box.
Turn it, the arcs and lozenges dissolve,
Spinning the good dreams and the humpbacked night
To overlapping folds of pink and violet,
The Rose window high in Lincoln's transept.
Granny, the frog, the little tiptoe train
Open and shut, the dog's blind, lustrous eyes
Bright as the eyes of spiderwebs and flowers,
As the next lucky thing hoped or counted for.
The magic lantern is hot silver tin
And if you touch it, it will burn your fingers.

The Master Builder

for Tommy Scupham

On leave, he must have slung his British Warm
Across this chair, hugged, kissed, and joshed them all –
That fire still warms the corners of the room.

His cap-badge showed the way the land could lie
For one more tommy coming home for good:
The Gunners' wheel above its *Ubique.*

He found the House that Jack built tumbling down
About their ears: the devil left to pay.
He picked things up; his ropes could take the strain.

He squared the family circle, found a skill
For throwing sky-hooks up against the clouds;
With blood for mortar raised the roof and wall

That set his brother and his sisters free.
They packed their cases, off to pack their brains
By ways that he had made. They made their way.

And where he grew, the little Market Town
Spun at his finger-tips: that tarnished wheel
Flashed out its golden spokes; became the sun.

A Monumental Mason. Near and wide
He played the jester, but they knew him king,
Bone-shaking off to Charlestons in his Ford,

Hanging one-handed, striking down a grin
From church-clock high: a monkey on a stick,
Or banging aces at some farmer's lawn.

He stood in his foundations, making clear
His rule, his measure. Children trooped the Yard,
The House that Jack built now set fair and square.

And still the family ghosts go whispering,
Whatever houses they inhabit now:
'You owe your brother Tommy everything.'

Teasing me still, he strolls the garden where
His old Victoria Plum nods in the wind,
And *Ubique*: his hands build everywhere.

Christmas 1987

Once more I cover this nakedness which is my own,
Beyond shame, wasting, wanting; soon to be ash,
Pulling her nightdress down over the sharp bone
Which bites through the shocked absences of flesh.
    I hold viburnum up to her face; it is the Host,
    The offer of wax to wax — I am becoming a ghost

As I fuss for the damp sheets rucking across the bed,
My breath crossed upon hers when the choked sighs come
And spoons of morphine swim to the propped head.
It is time to die in this house which was never home.
    Rough winter flowers grow slack against the light;
    The days are another way of spelling the night

In this molten place where a silver photograph-frame
Runs like mercury, shifting its beads about,
Rocking her once again in my father's dream.
In the long-ago garden, holding the summer out,
    Her eyes laugh for his love: the ridiculous, wide
    Hat over dark hair, the white dress poured into shade,

This house burning away in my arms: the flowers,
The pictures loose with their huge and radiant trees,
Their cornfields crushed and rippling to faraway spires,
The cards blurred with angels and unpacked memories,
    A letter alert for the quick ply, the thresh of life—
    I am cut between glittering knife and glittering knife,

All honed by dangerous love. I drift through signs,
Defenceless, for there is nothing left here to defend
But the wrapped gifts helpless in their tinsel chains.
Somewhere Christ has been born; here, the promised end
    Thirsts for that promise kept: her one desire
    The pure trouble of fire, and what will become of fire.

Young Ghost

Oh, the young ghost, her long hair coursing
Down to her shoulders: dark hair, the heat of the day
Sunk deep under those tucks and scents, drowsing
At the neck's nape — she looks so far away,
Though love twinned in her eyes has slipped its blindfold —

And really she glances across to him for ever,
His shutter chocking the light back into the box,
Snapping the catch on a purse of unchanged silver.
Under those seven seals and the seven locks
She is safe now from growing with what is growing,

And safe, too, from dying with what is dying,
Though her solemn flowers unpick themselves from her
  hands,
The dress rustles to moth-wings; her sweet flesh fraying
Out into knots and wrinkles and low-tide sands.
The hat is only a basket for thoughtless dust.

And she stands there lost in a smile in a black garden:
A white quotation floating away from its book.
Will it be silver, gold, or the plain-truth leaden?
But the camera chirps like a cricket, dies — and look,
She floats away light as ash in its tiny casket.

Dancing Shoes

At Time's *Excuse-me*, how could you refuse
A Quickstep on his wind-up gramophone?
How long since you wore out your dancing shoes,

Or his vest-pocket Kodak framed the views
In which you never found yourself alone?
At times, excuse me, how could you refuse

His Roaring Twenties, stepping out in twos,
Love singing in his lightest baritone?
How long since you wore out your dancing shoes

And shut away that music, hid the clues
By tennis-courts long rank and overgrown?
At times, excuse me, how could you refuse

To say his choice was just what you would choose,
Although he spun your fingers to the bone,
How long since you wore out your dancing shoes?

The Charleston and the Foxtrot and the Blues —
The records end in blur and monotone.
At Time's *Excuse-me* — how could you refuse?
How long since you wore out your dancing shoes?

Watching the Perseids: Remembering the Dead

The Perseids go riding softly down:
Hair-streak moths, brushing with faint wings
This audience of stars with sharp, young faces,
Staring our eyes out with such charming brilliance:
Life, set in its ways and constellations,
Which knows its magnitude, its name and status.

These, though, are whispered ones, looked for in August
Or when we trip on dead and dying birthdays,
Drinking a quiet toast at some green Christmas
To those, who, fallen from space and height
No longer reach us with their smoky fingers
Or touch this sheet of water under no moon.

They are the comet's tail we all must pass through
Dreamed out into a trail of Jack O'Lanterns,
A shattered windscreen on the road to nowhere.
We stand in this late dark-room, watch the Master
Swing his light-pencil, tentative yet certain,
As if calligraphy could tease out meaning,

And, between a huge water, huger sky,
Glimmers of something on the jimp horizon,
There might be pictures, might be conversations.
We wait for last words, ease the rites of passage,
The cold night hung in chains about our questions,
Our black ark swinging lightly to its mooring.

OXFORD POETS

Fleur Adcock
James Berry
Edward Kamau Brathwaite
Joseph Brodsky
Basil Bunting
W. H. Davies
Michael Donaghy
Keith Douglas
D. J. Enright
Roy Fisher
David Gascoyne
Ivor Gurney
David Harsent
Anthony Hecht
Zbigniew Herbert
Thomas Kinsella
Brad Leithauser
Derek Mahon
Medbh McGuckian
James Merrill
Peter Porter
Craig Raine
Christopher Reid
Stephen Romer
Carole Satyamurti
Peter Scupham
Penelope Shuttle
Louis Simpson
Anne Stevenson
George Szirtes
Grete Tartler
Edward Thomas
Anthony Thwaite
Charles Tomlinson
Chris Wallace-Crabbe
Hugo Williams